Mometrix
TEST PREPARATION

I0868212

Common Core High School: Algebra Secrets Study Guide

Dear Future Exam Success Story

First of all, **THANK YOU** for purchasing Mometrix study materials!

Second, congratulations! You are one of the few determined test-takers who are committed to doing whatever it takes to excel on your exam. **You have come to the right place.** We developed these study materials with one goal in mind: to deliver you the information you need in a format that's concise and easy to use.

In addition to optimizing your guide for the content of the test, we've outlined our recommended steps for breaking down the preparation process into small, attainable goals so you can make sure you stay on track.

We've also analyzed the entire test-taking process, identifying the most common pitfalls and showing how you can overcome them and be ready for any curveball the test throws you.

Standardized testing is one of the biggest obstacles on your road to success, which only increases the importance of doing well in the high-pressure, high-stakes environment of test day. Your results on this test could have a significant impact on your future, and this guide provides the information and practical advice to help you achieve your full potential on test day.

Your success is our success

We would love to hear from you! If you would like to share the story of your exam success or if you have any questions or comments in regard to our products, please contact us at **800-673-8175** or **support@mometrix.com**.

Thanks again for your business and we wish you continued success!

Sincerely,
The Mometrix Test Preparation Team

Need more help? Check out our flashcards at:
http://MometrixFlashcards.com/CCSS

TABLE OF CONTENTS

Introduction

Thank you for purchasing this resource! You have made the choice to prepare yourself for a test that could have a huge impact on your future, and this guide is designed to help you be fully ready for test day. Obviously, it's important to have a solid understanding of the test material, but you also need to be prepared for the unique environment and stressors of the test, so that you can perform to the best of your abilities.

For this purpose, the first section that appears in this guide is the **Secret Keys**. We've devoted countless hours to meticulously researching what works and what doesn't, and we've boiled down our findings to the five most impactful steps you can take to improve your performance on the test. We start at the beginning with study planning and move through the preparation process, all the way to the testing strategies that will help you get the most out of what you know when you're finally sitting in front of the test.

We recommend that you start preparing for your test as far in advance as possible. However, if you've bought this guide as a last-minute study resource and only have a few days before your test, we recommend that you skip over the first two Secret Keys since they address a long-term study plan.

If you struggle with **test anxiety**, we strongly encourage you to check out our recommendations for how you can overcome it. Test anxiety is a formidable foe, but it can be beaten, and we want to make sure you have the tools you need to defeat it.

Secret Key #1 – Plan Big, Study Small

There's a lot riding on your performance. If you want to ace this test, you're going to need to keep your skills sharp and the material fresh in your mind. You need a plan that lets you review everything you need to know while still fitting in your schedule. We'll break this strategy down into three categories.

Information Organization

Start with the information you already have: the official test outline. From this, you can make a complete list of all the concepts you need to cover before the test. Organize these concepts into groups that can be studied together, and create a list of any related vocabulary you need to learn so you can brush up on any difficult terms. You'll want to keep this vocabulary list handy once you actually start studying since you may need to add to it along the way.

Time Management

Once you have your set of study concepts, decide how to spread them out over the time you have left before the test. Break your study plan into small, clear goals so you have a manageable task for each day and know exactly what you're doing. Then just focus on one small step at a time. When you manage your time this way, you don't need to spend hours at a time studying. Studying a small block of content for a short period each day helps you retain information better and avoid stressing over how much you have left to do. You can relax knowing that you have a plan to cover everything in time. In order for this strategy to be effective though, you have to start studying early and stick to your schedule. Avoid the exhaustion and futility that comes from last-minute cramming!

Study Environment

The environment you study in has a big impact on your learning. Studying in a coffee shop, while probably more enjoyable, is not likely to be as fruitful as studying in a quiet room. It's important to keep distractions to a minimum. You're only planning to study for a short block of time, so make the most of it. Don't pause to check your phone or get up to find a snack. It's also important to **avoid multitasking**. Research has consistently shown that multitasking will make your studying dramatically less effective. Your study area should also be comfortable and well-lit so you don't have the distraction of straining your eyes or sitting on an uncomfortable chair.

The time of day you study is also important. You want to be rested and alert. Don't wait until just before bedtime. Study when you'll be most likely to comprehend and remember. Even better, if you know what time of day your test will be, set that time aside for study. That way your brain will be used to working on that subject at that specific time and you'll have a better chance of recalling information.

Finally, it can be helpful to team up with others who are studying for the same test. Your actual studying should be done in as isolated an environment as possible, but the work of organizing the information and setting up the study plan can be divided up. In between study sessions, you can discuss with your teammates the concepts that you're all studying and quiz each other on the details. Just be sure that your teammates are as serious about the test as you are. If you find that your study time is being replaced with social time, you might need to find a new team.

Secret Key #2 – Make Your Studying Count

You're devoting a lot of time and effort to preparing for this test, so you want to be absolutely certain it will pay off. This means doing more than just reading the content and hoping you can remember it on test day. It's important to make every minute of study count. There are two main areas you can focus on to make your studying count.

Retention

It doesn't matter how much time you study if you can't remember the material. You need to make sure you are retaining the concepts. To check your retention of the information you're learning, try recalling it at later times with minimal prompting. Try carrying around flashcards and glance at one or two from time to time or ask a friend who's also studying for the test to quiz you.

To enhance your retention, look for ways to put the information into practice so that you can apply it rather than simply recalling it. If you're using the information in practical ways, it will be much easier to remember. Similarly, it helps to solidify a concept in your mind if you're not only reading it to yourself but also explaining it to someone else. Ask a friend to let you teach them about a concept you're a little shaky on (or speak aloud to an imaginary audience if necessary). As you try to summarize, define, give examples, and answer your friend's questions, you'll understand the concepts better and they will stay with you longer. Finally, step back for a big picture view and ask yourself how each piece of information fits with the whole subject. When you link the different concepts together and see them working together as a whole, it's easier to remember the individual components.

Finally, practice showing your work on any multi-step problems, even if you're just studying. Writing out each step you take to solve a problem will help solidify the process in your mind, and you'll be more likely to remember it during the test.

Modality

Modality simply refers to the means or method by which you study. Choosing a study modality that fits your own individual learning style is crucial. No two people learn best in exactly the same way, so it's important to know your strengths and use them to your advantage.

For example, if you learn best by visualization, focus on visualizing a concept in your mind and draw an image or a diagram. Try color-coding your notes, illustrating them, or creating symbols that will trigger your mind to recall a learned concept. If you learn best by hearing or discussing information, find a study partner who learns the same way or read aloud to yourself. Think about how to put the information in your own words. Imagine that you are giving a lecture on the topic and record yourself so you can listen to it later.

For any learning style, flashcards can be helpful. Organize the information so you can take advantage of spare moments to review. Underline key words or phrases. Use different colors for different categories. Mnemonic devices (such as creating a short list in which every item starts with the same letter) can also help with retention. Find what works best for you and use it to store the information in your mind most effectively and easily.

Secret Key #3 – Practice the Right Way

Your success on test day depends not only on how many hours you put into preparing, but also on whether you prepared the right way. It's good to check along the way to see if your studying is paying off. One of the most effective ways to do this is by taking practice tests to evaluate your progress. Practice tests are useful because they show exactly where you need to improve. Every time you take a practice test, pay special attention to these three groups of questions:

- The questions you got wrong
- The questions you had to guess on, even if you guessed right
- The questions you found difficult or slow to work through

This will show you exactly what your weak areas are, and where you need to devote more study time. Ask yourself why each of these questions gave you trouble. Was it because you didn't understand the material? Was it because you didn't remember the vocabulary? Do you need more repetitions on this type of question to build speed and confidence? Dig into those questions and figure out how you can strengthen your weak areas as you go back to review the material.

Additionally, many practice tests have a section explaining the answer choices. It can be tempting to read the explanation and think that you now have a good understanding of the concept. However, an explanation likely only covers part of the question's broader context. Even if the explanation makes perfect sense, **go back and investigate** every concept related to the question until you're positive you have a thorough understanding.

As you go along, keep in mind that the practice test is just that: practice. Memorizing these questions and answers will not be very helpful on the actual test because it is unlikely to have any of the same exact questions. If you only know the right answers to the sample questions, you won't be prepared for the real thing. **Study the concepts** until you understand them fully, and then you'll be able to answer any question that shows up on the test.

It's important to wait on the practice tests until you're ready. If you take a test on your first day of study, you may be overwhelmed by the amount of material covered and how much you need to learn. Work up to it gradually.

On test day, you'll need to be prepared for answering questions, managing your time, and using the test-taking strategies you've learned. It's a lot to balance, like a mental marathon that will have a big impact on your future. Like training for a marathon, you'll need to start slowly and work your way up. When test day arrives, you'll be ready.

Start with the strategies you've read in the first two Secret Keys—plan your course and study in the way that works best for you. If you have time, consider using multiple study resources to get different approaches to the same concepts. It can be helpful to see difficult concepts from more than one angle. Then find a good source for practice tests. Many times, the test website will suggest potential study resources or provide sample tests.

Practice Test Strategy

If you're able to find at least three practice tests, we recommend this strategy:

Untimed and Open-Book Practice

Take the first test with no time constraints and with your notes and study guide handy. Take your time and focus on applying the strategies you've learned.

Timed and Open-Book Practice

Take the second practice test open-book as well, but set a timer and practice pacing yourself to finish in time.

Timed and Closed-Book Practice

Take any other practice tests as if it were test day. Set a timer and put away your study materials. Sit at a table or desk in a quiet room, imagine yourself at the testing center, and answer questions as quickly and accurately as possible.

Keep repeating timed and closed-book tests on a regular basis until you run out of practice tests or it's time for the actual test. Your mind will be ready for the schedule and stress of test day, and you'll be able to focus on recalling the material you've learned.

Secret Key #4 – Pace Yourself

Once you're fully prepared for the material on the test, your biggest challenge on test day will be managing your time. Just knowing that the clock is ticking can make you panic even if you have plenty of time left. Work on pacing yourself so you can build confidence against the time constraints of the exam. Pacing is a difficult skill to master, especially in a high-pressure environment, so **practice is vital**.

Set time expectations for your pace based on how much time is available. For example, if a section has 60 questions and the time limit is 30 minutes, you know you have to average 30 seconds or less per question in order to answer them all. Although 30 seconds is the hard limit, set 25 seconds per question as your goal, so you reserve extra time to spend on harder questions. When you budget extra time for the harder questions, you no longer have any reason to stress when those questions take longer to answer.

Don't let this time expectation distract you from working through the test at a calm, steady pace, but keep it in mind so you don't spend too much time on any one question. Recognize that taking extra time on one question you don't understand may keep you from answering two that you do understand later in the test. If your time limit for a question is up and you're still not sure of the answer, mark it and move on, and come back to it later if the time and the test format allow. If the testing format doesn't allow you to return to earlier questions, just make an educated guess; then put it out of your mind and move on.

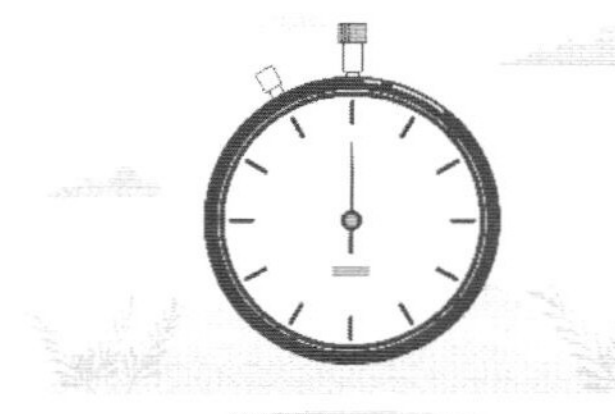

On the easier questions, be careful not to rush. It may seem wise to hurry through them so you have more time for the challenging ones, but it's not worth missing one if you know the concept and just didn't take the time to read the question fully. Work efficiently but make sure you understand the question and have looked at all of the answer choices, since more than one may seem right at first.

Even if you're paying attention to the time, you may find yourself a little behind at some point. You should speed up to get back on track, but do so wisely. Don't panic; just take a few seconds less on each question until you're caught up. Don't guess without thinking, but do look through the answer choices and eliminate any you know are wrong. If you can get down to two choices, it is often worthwhile to guess from those. Once you've chosen an answer, move on and don't dwell on any that you skipped or had to hurry through. If a question was taking too long, chances are it was one of the harder ones, so you weren't as likely to get it right anyway.

On the other hand, if you find yourself getting ahead of schedule, it may be beneficial to slow down a little. The more quickly you work, the more likely you are to make a careless mistake that will affect your score. You've budgeted time for each question, so don't be afraid to spend that time. Practice an efficient but careful pace to get the most out of the time you have.

Secret Key #5 – Have a Plan for Guessing

When you're taking the test, you may find yourself stuck on a question. Some of the answer choices seem better than others, but you don't see the one answer choice that is obviously correct. What do you do?

The scenario described above is very common, yet most test takers have not effectively prepared for it. Developing and practicing a plan for guessing may be one of the single most effective uses of your time as you get ready for the exam.

In developing your plan for guessing, there are three questions to address:

- When should you start the guessing process?
- How should you narrow down the choices?
- Which answer should you choose?

When to Start the Guessing Process

Unless your plan for guessing is to select C every time (which, despite its merits, is not what we recommend), you need to leave yourself enough time to apply your answer elimination strategies. Since you have a limited amount of time for each question, that means that if you're going to give yourself the best shot at guessing correctly, you have to decide quickly whether or not you will guess.

Of course, the best-case scenario is that you don't have to guess at all, so first, see if you can answer the question based on your knowledge of the subject and basic reasoning skills. Focus on the key words in the question and try to jog your memory of related topics. Give yourself a chance to bring the knowledge to mind, but once you realize that you don't have (or you can't access) the knowledge you need to answer the question, it's time to start the guessing process.

It's almost always better to start the guessing process too early than too late. It only takes a few seconds to remember something and answer the question from knowledge. Carefully eliminating wrong answer choices takes longer. Plus, going through the process of eliminating answer choices can actually help jog your memory.

Summary: Start the guessing process as soon as you decide that you can't answer the question based on your knowledge.

How to Narrow Down the Choices

The next chapter in this book (**Test-Taking Strategies**) includes a wide range of strategies for how to approach questions and how to look for answer choices to eliminate. You will definitely want to read those carefully, practice them, and figure out which ones work best for you. Here though, we're going to address a mindset rather than a particular strategy.

Your odds of guessing an answer correctly depend on how many options you are choosing from.

Number of options left	5	4	3	2	1
Odds of guessing correctly	20%	25%	33%	50%	100%

You can see from this chart just how valuable it is to be able to eliminate incorrect answers and make an educated guess, but there are two things that many test takers do that cause them to miss out on the benefits of guessing:

- Accidentally eliminating the correct answer
- Selecting an answer based on an impression

We'll look at the first one here, and the second one in the next section.

To avoid accidentally eliminating the correct answer, we recommend a thought exercise called **the $5 challenge**. In this challenge, you only eliminate an answer choice from contention if you are willing to bet $5 on it being wrong. Why $5? Five dollars is a small but not insignificant amount of money. It's an amount you could afford to lose but wouldn't want to throw away. And while losing $5 once might not hurt too much, doing it twenty times will set you back $100. In the same way, each small decision you make—eliminating a choice here, guessing on a question there—won't by itself impact your score very much, but when you put them all together, they can make a big difference. By holding each answer choice elimination decision to a higher standard, you can reduce the risk of accidentally eliminating the correct answer.

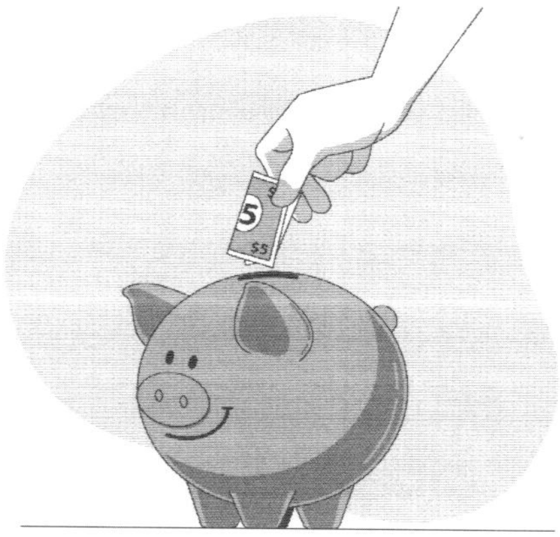

The $5 challenge can also be applied in a positive sense: If you are willing to bet $5 that an answer choice *is* correct, go ahead and mark it as correct.

Summary: Only eliminate an answer choice if you are willing to bet $5 that it is wrong.

Which Answer to Choose

You're taking the test. You've run into a hard question and decided you'll have to guess. You've eliminated all the answer choices you're willing to bet $5 on. Now you have to pick an answer. Why do we even need to talk about this? Why can't you just pick whichever one you feel like when the time comes?

The answer to these questions is that if you don't come into the test with a plan, you'll rely on your impression to select an answer choice, and if you do that, you risk falling into a trap. The test writers know that everyone who takes their test will be guessing on some of the questions, so they intentionally write wrong answer choices to seem plausible. You still have to pick an answer though, and if the wrong answer choices are designed to look right, how can you ever be sure that you're not falling for their trap? The best solution we've found to this dilemma is to take the decision out of your hands entirely. Here is the process we recommend:

Once you've eliminated any choices that you are confident (willing to bet $5) are wrong, select the first remaining choice as your answer.

Whether you choose to select the first remaining choice, the second, or the last, the important thing is that you use some preselected standard. Using this approach guarantees that you will not be enticed into selecting an answer choice that looks right, because you are not basing your decision on how the answer choices look.

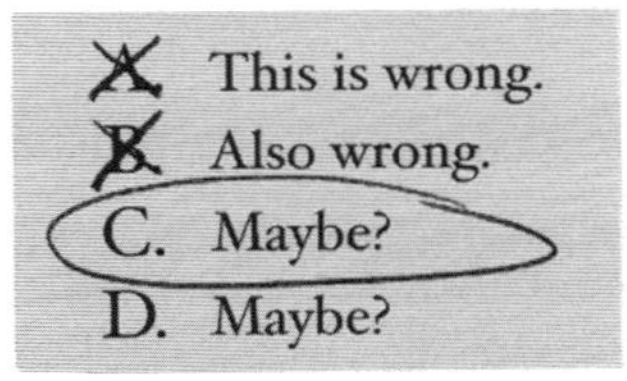

This is not meant to make you question your knowledge. Instead, it is to help you recognize the difference between your knowledge and your impressions. There's a huge difference between thinking an answer is right because of what you know, and thinking an answer is right because it looks or sounds like it should be right.

Summary: To ensure that your selection is appropriately random, make a predetermined selection from among all answer choices you have not eliminated.

Test-Taking Strategies

This section contains a list of test-taking strategies that you may find helpful as you work through the test. By taking what you know and applying logical thought, you can maximize your chances of answering any question correctly!

It is very important to realize that every question is different and every person is different: no single strategy will work on every question, and no single strategy will work for every person. That's why we've included all of them here, so you can try them out and determine which ones work best for different types of questions and which ones work best for you.

Question Strategies

☑ Read Carefully

Read the question and the answer choices carefully. Don't miss the question because you misread the terms. You have plenty of time to read each question thoroughly and make sure you understand what is being asked. Yet a happy medium must be attained, so don't waste too much time. You must read carefully and efficiently.

☑ Contextual Clues

Look for contextual clues. If the question includes a word you are not familiar with, look at the immediate context for some indication of what the word might mean. Contextual clues can often give you all the information you need to decipher the meaning of an unfamiliar word. Even if you can't determine the meaning, you may be able to narrow down the possibilities enough to make a solid guess at the answer to the question.

☑ Prefixes

If you're having trouble with a word in the question or answer choices, try dissecting it. Take advantage of every clue that the word might include. Prefixes can be a huge help. Usually, they allow you to determine a basic meaning. *Pre-* means before, *post-* means after, *pro-* is positive, *de-* is negative. From prefixes, you can get an idea of the general meaning of the word and try to put it into context.

☑ Hedge Words

Watch out for critical hedge words, such as *likely, may, can, often, almost, mostly, usually, generally, rarely,* and *sometimes.* Question writers insert these hedge phrases to cover every possibility. Often an answer choice will be wrong simply because it leaves no room for exception. Be on guard for answer choices that have definitive words such as *exactly* and *always.*

☑ Switchback Words

Stay alert for *switchbacks*. These are the words and phrases frequently used to alert you to shifts in thought. The most common switchback words are *but, although*, and *however*. Others include *nevertheless, on the other hand, even though, while, in spite of, despite*, and *regardless of.* Switchback words are important to catch because they can change the direction of the question or an answer choice.

⊘ FACE VALUE

When in doubt, use common sense. Accept the situation in the problem at face value. Don't read too much into it. These problems will not require you to make wild assumptions. If you have to go beyond creativity and warp time or space in order to have an answer choice fit the question, then you should move on and consider the other answer choices. These are normal problems rooted in reality. The applicable relationship or explanation may not be readily apparent, but it is there for you to figure out. Use your common sense to interpret anything that isn't clear.

Answer Choice Strategies

⊘ ANSWER SELECTION

The most thorough way to pick an answer choice is to identify and eliminate wrong answers until only one is left, then confirm it is the correct answer. Sometimes an answer choice may immediately seem right, but be careful. The test writers will usually put more than one reasonable answer choice on each question, so take a second to read all of them and make sure that the other choices are not equally obvious. As long as you have time left, it is better to read every answer choice than to pick the first one that looks right without checking the others.

⊘ ANSWER CHOICE FAMILIES

An answer choice family consists of two (in rare cases, three) answer choices that are very similar in construction and cannot all be true at the same time. If you see two answer choices that are direct opposites or parallels, one of them is usually the correct answer. For instance, if one answer choice says that quantity *x* increases and another either says that quantity *x* decreases (opposite) or says that quantity *y* increases (parallel), then those answer choices would fall into the same family. An answer choice that doesn't match the construction of the answer choice family is more likely to be incorrect. Most questions will not have answer choice families, but when they do appear, you should be prepared to recognize them.

⊘ ELIMINATE ANSWERS

Eliminate answer choices as soon as you realize they are wrong, but make sure you consider all possibilities. If you are eliminating answer choices and realize that the last one you are left with is also wrong, don't panic. Start over and consider each choice again. There may be something you missed the first time that you will realize on the second pass.

⊘ AVOID FACT TRAPS

Don't be distracted by an answer choice that is factually true but doesn't answer the question. You are looking for the choice that answers the question. Stay focused on what the question is asking for so you don't accidentally pick an answer that is true but incorrect. Always go back to the question and make sure the answer choice you've selected actually answers the question and is not merely a true statement.

⊘ EXTREME STATEMENTS

In general, you should avoid answers that put forth extreme actions as standard practice or proclaim controversial ideas as established fact. An answer choice that states the "process should be used in certain situations, if…" is much more likely to be correct than one that states the "process should be discontinued completely." The first is a calm rational statement and doesn't even make a definitive, uncompromising stance, using a hedge word *if* to provide wiggle room, whereas the second choice is far more extreme.

⊘ BENCHMARK

As you read through the answer choices and you come across one that seems to answer the question well, mentally select that answer choice. This is not your final answer, but it's the one that will help you evaluate the other answer choices. The one that you selected is your benchmark or standard for judging each of the other answer choices. Every other answer choice must be compared to your benchmark. That choice is correct until proven otherwise by another answer choice beating it. If you find a better answer, then that one becomes your new benchmark. Once you've decided that no other choice answers the question as well as your benchmark, you have your final answer.

⊘ PREDICT THE ANSWER

Before you even start looking at the answer choices, it is often best to try to predict the answer. When you come up with the answer on your own, it is easier to avoid distractions and traps because you will know exactly what to look for. The right answer choice is unlikely to be word-for-word what you came up with, but it should be a close match. Even if you are confident that you have the right answer, you should still take the time to read each option before moving on.

General Strategies

⊘ TOUGH QUESTIONS

If you are stumped on a problem or it appears too hard or too difficult, don't waste time. Move on! Remember though, if you can quickly check for obviously incorrect answer choices, your chances of guessing correctly are greatly improved. Before you completely give up, at least try to knock out a couple of possible answers. Eliminate what you can and then guess at the remaining answer choices before moving on.

⊘ CHECK YOUR WORK

Since you will probably not know every term listed and the answer to every question, it is important that you get credit for the ones that you do know. Don't miss any questions through careless mistakes. If at all possible, try to take a second to look back over your answer selection and make sure you've selected the correct answer choice and haven't made a costly careless mistake (such as marking an answer choice that you didn't mean to mark). This quick double check should more than pay for itself in caught mistakes for the time it costs.

⊘ PACE YOURSELF

It's easy to be overwhelmed when you're looking at a page full of questions; your mind is confused and full of random thoughts, and the clock is ticking down faster than you would like. Calm down and maintain the pace that you have set for yourself. Especially as you get down to the last few minutes of the test, don't let the small numbers on the clock make you panic. As long as you are on track by monitoring your pace, you are guaranteed to have time for each question.

⊘ DON'T RUSH

It is very easy to make errors when you are in a hurry. Maintaining a fast pace in answering questions is pointless if it makes you miss questions that you would have gotten right otherwise. Test writers like to include distracting information and wrong answers that seem right. Taking a little extra time to avoid careless mistakes can make all the difference in your test score. Find a pace that allows you to be confident in the answers that you select.

⊘ Keep Moving

Panicking will not help you pass the test, so do your best to stay calm and keep moving. Taking deep breaths and going through the answer elimination steps you practiced can help to break through a stress barrier and keep your pace.

Final Notes

The combination of a solid foundation of content knowledge and the confidence that comes from practicing your plan for applying that knowledge is the key to maximizing your performance on test day. As your foundation of content knowledge is built up and strengthened, you'll find that the strategies included in this chapter become more and more effective in helping you quickly sift through the distractions and traps of the test to isolate the correct answer.

Now that you're preparing to move forward into the test content chapters of this book, be sure to keep your goal in mind. As you read, think about how you will be able to apply this information on the test. If you've already seen sample questions for the test and you have an idea of the question format and style, try to come up with questions of your own that you can answer based on what you're reading. This will give you valuable practice applying your knowledge in the same ways you can expect to on test day.

Good luck and good studying!

Number and Quantity

Number Basics

Classifications of Numbers

Numbers are the basic building blocks of mathematics. Specific features of numbers are identified by the following terms:

Integer – any positive or negative whole number, including zero. Integers do not include fractions $\left(\frac{1}{3}\right)$, decimals (0.56), or mixed numbers $\left(7\frac{3}{4}\right)$.

Prime number – any whole number greater than 1 that has only two factors, itself and 1; that is, a number that can be divided evenly only by 1 and itself.

Composite number – any whole number greater than 1 that has more than two different factors; in other words, any whole number that is not a prime number. For example: The composite number 8 has the factors of 1, 2, 4, and 8.

Even number – any integer that can be divided by 2 without leaving a remainder. For example: 2, 4, 6, 8, and so on.

Odd number – any integer that cannot be divided evenly by 2. For example: 3, 5, 7, 9, and so on.

Decimal number – any number that uses a decimal point to show the part of the number that is less than one. Example: 1.234.

Decimal point – a symbol used to separate the ones place from the tenths place in decimals or dollars from cents in currency.

Decimal place – the position of a number to the right of the decimal point. In the decimal 0.123, the 1 is in the first place to the right of the decimal point, indicating tenths; the 2 is in the second place, indicating hundredths; and the 3 is in the third place, indicating thousandths.

The **decimal**, or base 10, system is a number system that uses ten different digits (0, 1, 2, 3, 4, 5, 6, 7, 8, 9). An example of a number system that uses something other than ten digits is the **binary**, or base 2, number system, used by computers, which uses only the numbers 0 and 1. It is thought that the decimal system originated because people had only their 10 fingers for counting.

Rational numbers include all integers, decimals, and fractions. Any terminating or repeating decimal number is a rational number.

Irrational numbers cannot be written as fractions or decimals because the number of decimal places is infinite and there is no recurring pattern of digits within the number. For example, pi (π) begins with 3.141592 and continues without terminating or repeating, so pi is an irrational number.

Real numbers are the set of all rational and irrational numbers.

Review Video: Classification of Numbers
Visit mometrix.com/academy and enter code: 461071

Review Video: Prime and Composite Numbers
Visit mometrix.com/academy and enter code: 565581

Numbers in Word Form and Place Value

When writing numbers out in word form or translating word form to numbers, it is essential to understand how a place value system works. In the decimal or base-10 system, each digit of a number represents how many of the corresponding place value—a specific factor of 10—are contained in the number being represented. To make reading numbers easier, every three digits to the left of the decimal place is preceded by a comma. The following table demonstrates some of the place values:

Power of 10	10^3	10^2	10^1	10^0	10^{-1}	10^{-2}	10^{-3}
Value	1,000	100	10	1	0.1	0.01	0.001
Place	thousands	hundreds	tens	ones	tenths	hundredths	thousandths

For example, consider the number 4,546.09, which can be separated into each place value like this:

4: thousands
5: hundreds
4: tens
6: ones
0: tenths
9: hundredths

This number in word form would be *four thousand five hundred forty-six and nine hundredths.*

Review Video: Place Value
Visit mometrix.com/academy and enter code: 205433

Number Lines

A number line is a graph to see the distance between numbers. Basically, this graph shows the relationship between numbers. So a number line may have a point for zero and may show negative numbers on the left side of the line. Any positive numbers are placed on the right side of the line. For example, consider the points labeled on the following number line:

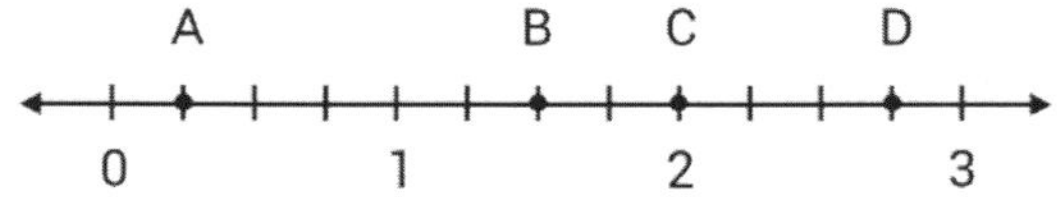

We can use the dashed lines on the number line to identify each point. Each dashed line between two whole numbers is $\frac{1}{4}$. The line halfway between two numbers is $\frac{1}{2}$.

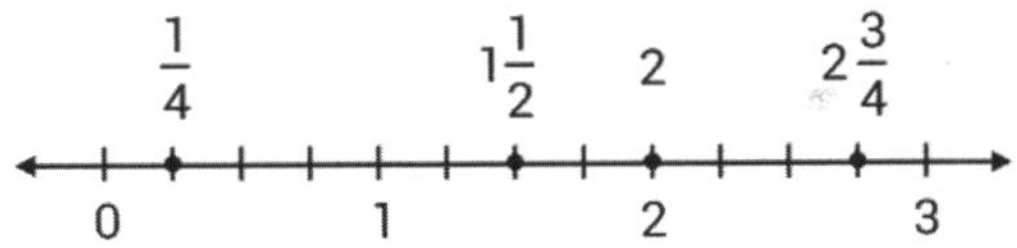

Review Video: The Number Line
Visit mometrix.com/academy and enter code: 816439

Comparing Numbers

Inequality Notation

The symbols $<$ and $>$ mean "is less than" and "is greater than," respectively. For instance, $3 < 5$ means "3 is less than 5," and $7 > 4$ means "7 is greater than 4." Statements like $3 < 5$ and $7 > 4$ are **inequalities**, and the symbols $<$ and $>$ are **inequality symbols**.

Whole Numbers and Decimal Numbers

To compare whole or decimal numbers, we look at the most significant place (the leftmost digit) at which they differ. The number with the larger digit in that place is larger. For instance, $0.3\underline{8}74$ and $0.3\underline{9}$ differ in the hundredths place (underlined). Since 8 is smaller than 9, we see $0.3874 < 0.39$. This is clearer if we make the decimals equal in length by writing extra zeroes: $0.3874 < 0.3900$. Similarly, $2\underline{3}.984 < 2\underline{5}.112$ because 3 is smaller than 5, or 23 is smaller than 25.

Fractions

If fractions have the same denominator, the fraction with the larger numerator is larger. For instance, $\frac{2}{7} < \frac{5}{7}$ since $2 < 5$. We compare fractions with different denominators by finding a common denominator. When comparing the fractions with a common denominator we only compare the numerator, so as a shortcut, we can multiply each numerator by the denominator of the other fraction. The numerator that produces the larger product belongs to the larger fraction. For example, to compare $\frac{7}{8}$ and $\frac{5}{6}$, we note that $7 \cdot 6 = 42$ is larger than $5 \cdot 8 = 40$. Since the numerator 7 produces the larger product, we see $\frac{7}{8} > \frac{5}{6}$. We can also compare fractions by converting them to decimals. For instance, since $\frac{3}{4} = 0.75$ and $\frac{4}{5} = 0.8$ and $0.75 < 0.8$, we conclude $\frac{3}{4} < \frac{4}{5}$.

Mixed Numbers

To compare mixed numbers we compare their whole number parts. If those are equal, then we compare their fractional parts. For instance, $5\frac{3}{8} > 4\frac{7}{8}$ because $5 > 4$, but $3\frac{5}{9} < 3\frac{8}{9}$ because $\frac{5}{9} < \frac{8}{9}$.

Square Roots

To compare square roots, we convert them to decimals, usually with a calculator. To compare two square roots, we compare their radicands. For instance, $\sqrt{11} < \sqrt{14}$ because $11 < 14$.

Negative Numbers

A negative number is always less than a positive number. Two negative numbers compare in the reverse order of their opposites. For instance, $-6 < -2$ (that is, -6 is smaller, more negative, than -2) because $6 > 2$.

Absolute Value

A precursor to working with negative numbers is understanding what **absolute values** are. A number's absolute value is simply the distance away from zero a number is on the number line. The absolute value of a number is always positive and is written $|x|$. For example, the absolute value of 3, written as $|3|$, is 3 because the distance between 0 and 3 on a number line is three units. Likewise, the absolute value of -3, written as $|-3|$, is 3 because the distance between 0 and -3 on a number line is three units. So $|3| = |-3|$.

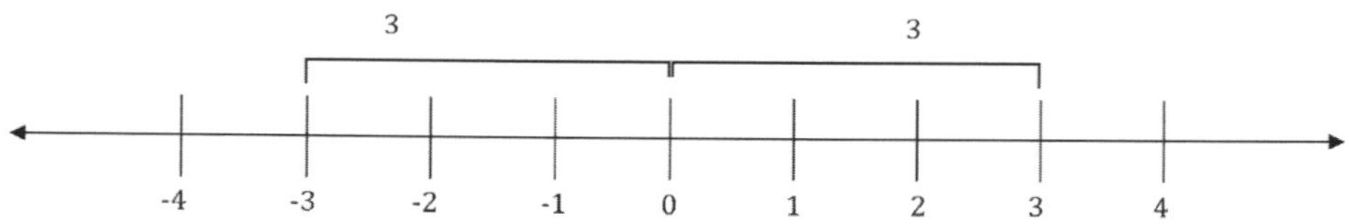

Review Video: Absolute Value
Visit mometrix.com/academy and enter code: 314669

Operations

An **operation** is simply a mathematical process that takes some value(s) as input(s) and produces an output. Elementary operations are often written in the following form: *value operation value*. For instance, in the expression $1 + 2$ the values are 1 and 2 and the operation is addition. Performing the operation gives the output of 3. In this way we can say that $1 + 2$ and 3 are equal, or $1 + 2 = 3$.

Addition

Addition increases the value of one quantity by the value of another quantity (both called **addends**). Example: $2 + 4 = 6$ or $8 + 9 = 17$. The result is called the **sum**. With addition, the order does not matter, $4 + 2 = 2 + 4$.

When adding signed numbers, if the signs are the same simply add the absolute values of the addends and apply the original sign to the sum. For example, $(+4) + (+8) = +12$ and $(-4) + (-8) = -12$. When the original signs are different, take the absolute values of the addends and subtract the smaller value from the larger value, then apply the original sign of the larger value to the difference. Example: $(+4) + (-8) = -4$ and $(-4) + (+8) = +4$.

Subtraction

Subtraction is the opposite operation to addition; it decreases the value of one quantity (the **minuend**) by the value of another quantity (the **subtrahend**). For example, $6 - 4 = 2$ or $17 - 8 = 9$. The result is called the **difference**. Note that with subtraction, the order does matter, $6 - 4 \neq 4 - 6$.

For subtracting signed numbers, change the sign of the subtrahend and then follow the same rules used for addition. Example: $(+4) - (+8) = (+4) + (-8) = -4$

Multiplication

Multiplication can be thought of as repeated addition. One number (the **multiplier**) indicates how many times to add the other number (the **multiplicand**) to itself. Example: $3 \times 2 = 2 + 2 + 2 = 6$. With multiplication, the order does not matter, $2 \times 3 = 3 \times 2$ or $3 + 3 = 2 + 2 + 2$, either way the result (the **product**) is the same.

If the signs are the same, the product is positive when multiplying signed numbers. Example: $(+4) \times (+8) = +32$ and $(-4) \times (-8) = +32$. If the signs are opposite, the product is negative. Example: $(+4) \times (-8) = -32$ and $(-4) \times (+8) = -32$. When more than two factors are multiplied together, the sign of the product is determined by how many negative factors are present. If there are an odd number of negative factors then the product is negative, whereas an even number of negative factors indicates a positive product. Example: $(+4) \times (-8) \times (-2) = +64$ and $(-4) \times (-8) \times (-2) = -64$.

Division

Division is the opposite operation to multiplication; one number (the **divisor**) tells us how many parts to divide the other number (the **dividend**) into. The result of division is called the **quotient**. Example: $20 \div 4 = 5$. If 20 is split into 4 equal parts, each part is 5. With division, the order of the numbers does matter, $20 \div 4 \neq 4 \div 20$.

The rules for dividing signed numbers are similar to multiplying signed numbers. If the dividend and divisor have the same sign, the quotient is positive. If the dividend and divisor have opposite signs, the quotient is negative. Example: $(-4) \div (+8) = -0.5$.

Review Video: Mathematical Operations
Visit mometrix.com/academy and enter code: 208095

Parentheses

Parentheses are used to designate which operations should be done first when there are multiple operations. Example: $4 - (2 + 1) = 1$; the parentheses tell us that we must add 2 and 1, and then subtract the sum from 4, rather than subtracting 2 from 4 and then adding 1 (this would give us an answer of 3).

Review Video: Mathematical Parentheses
Visit mometrix.com/academy and enter code: 978600

Exponents

An **exponent** is a superscript number placed next to another number at the top right. It indicates how many times the base number is to be multiplied by itself. Exponents provide a shorthand way to write what would be a longer mathematical expression. Example: $2^4 = 2 \times 2 \times 2 \times 2$. A number with an exponent of 2 is said to be "squared," while a number with an exponent of 3 is said to be "cubed." The value of a number raised to an exponent is called its power. So 8^4 is read as "8 to the 4th power," or "8 raised to the power of 4."

Review Video: Exponents
Visit mometrix.com/academy and enter code: 600998

Roots

A **root**, such as a square root, is another way of writing a fractional exponent. Instead of using a superscript, roots use the radical symbol ($\sqrt{\ }$) to indicate the operation. A radical will have a number underneath the bar, and may sometimes have a number in the upper left: $\sqrt[n]{a}$, read as "the nth root of a." The relationship between radical notation and exponent notation can be described by this equation:

$$\sqrt[n]{a} = a^{\frac{1}{n}}$$

The two special cases of $n = 2$ and $n = 3$ are called square roots and cube roots. If there is no number to the upper left, the radical is understood to be a square root ($n = 2$). Nearly all of the roots you encounter will be square roots. A square root is the same as a number raised to the one-half power. When we say that a is the square root of b ($a = \sqrt{b}$), we mean that a multiplied by itself equals b: ($a \times a = b$).

A **perfect square** is a number that has an integer for its square root. There are 10 perfect squares from 1 to 100: 1, 4, 9, 16, 25, 36, 49, 64, 81, 100 (the squares of integers 1 through 10).

Review Video: Roots
Visit mometrix.com/academy and enter code: 795655

Review Video: Perfect Squares and Square Roots
Visit mometrix.com/academy and enter code: 648063

Word Problems and Mathematical Symbols

When working on word problems, you must be able to translate verbal expressions or "math words" into math symbols. This chart contains several "math words" and their appropriate symbols:

Phrase	Symbol
equal, is, was, will be, has, costs, gets to, is the same as, becomes	$=$
times, of, multiplied by, product of, twice, doubles, halves, triples	$\times$
divided by, per, ratio of/to, out of	$\div$
plus, added to, sum, combined, and, more than, totals of	$+$
subtracted from, less than, decreased by, minus, difference between	$-$
what, how much, original value, how many, a number, a variable	x, n, etc.

Review Video: Understanding Word Problems
Visit mometrix.com/academy and enter code: 499199

Examples of Translated Mathematical Phrases

- The phrase four more than twice a number can be written algebraically as $2x + 4$.
- The phrase half a number decreased by six can be written algebraically as $\frac{1}{2}x - 6$.
- The phrase the sum of a number and the product of five and that number can be written algebraically as $x + 5x$.

- You may see a test question that says, "Olivia is constructing a bookcase from seven boards. Two of them are for vertical supports and five are for shelves. The height of the bookcase is twice the width of the bookcase. If the seven boards total 36 feet in length, what will be the height of Olivia's bookcase?" You would need to make a sketch and then create the equation to determine the width of the shelves. The height can be represented as double the width. (If x represents the width of the shelves in feet, then the height of the bookcase is $2x$. Since the seven boards total 36 feet, $2x + 2x + x + x + x + x + x = 36$ or $9x = 36$; $x = 4$. The height is twice the width, or 8 feet.)

Subtraction with Regrouping

A great way to make use of some of the features built into the decimal system would be regrouping when attempting longform subtraction operations. When subtracting within a place value, sometimes the minuend is smaller than the subtrahend; **regrouping** enables you to 'borrow' a unit from a place value to the left in order to get a positive difference. For example, consider subtracting 189 from 525 with regrouping.

First, set up the subtraction problem in vertical form:

$$\begin{array}{r} 525 \\ -\ 189 \\ \hline \end{array}$$

Notice that the numbers in the ones and tens columns of 525 are smaller than the numbers in the ones and tens columns of 189. This means you will need to use regrouping to perform subtraction:

$$\begin{array}{cccc} & 5 & 2 & 5 \\ - & 1 & 8 & 9 \\ \hline \end{array}$$

To subtract 9 from 5 in the ones column you will need to borrow from the 2 in the tens column:

$$\begin{array}{cccc} & 5 & 1 & 15 \\ - & 1 & 8 & 9 \\ \hline & & & 6 \end{array}$$

Next, to subtract 8 from 1 in the tens column you will need to borrow from the 5 in the hundreds column:

$$\begin{array}{cccc} & 4 & 11 & 15 \\ - & 1 & 8 & 9 \\ \hline & & 3 & 6 \end{array}$$

Last, subtract the 1 from the 4 in the hundreds column:

$$\begin{array}{cccc} & 4 & 11 & 15 \\ - & 1 & 8 & 9 \\ \hline & 3 & 3 & 6 \end{array}$$

Review Video: Subtracting Large Numbers
Visit mometrix.com/academy and enter code: 603350

Order of Operations

The **order of operations** is a set of rules that dictates the order in which we must perform each operation in an expression so that we will evaluate it accurately. If we have an expression that includes multiple different operations, the order of operations tells us which operations to do first. The most common mnemonic for the order of operations is **PEMDAS**, or "Please Excuse My Dear Aunt Sally." PEMDAS stands for parentheses, exponents, multiplication, division, addition, and subtraction. It is important to understand that multiplication and division have equal precedence, as do addition and subtraction, so those pairs of operations are simply worked from left to right in order.

For example, evaluating the expression $5 + 20 \div 4 \times (2 + 3)^2 - 6$ using the correct order of operations would be done like this:

- **P:** Perform the operations inside the parentheses: $(2 + 3) = 5$
- **E:** Simplify the exponents: $(5)^2 = 5 \times 5 = 25$
  - The expression now looks like this: $5 + 20 \div 4 \times 25 - 6$
- **MD:** Perform multiplication and division from left to right: $20 \div 4 = 5$; then $5 \times 25 = 125$
  - The expression now looks like this: $5 + 125 - 6$
- **AS:** Perform addition and subtraction from left to right: $5 + 125 = 130$; then $130 - 6 = 124$

Review Video: Order of Operations
Visit mometrix.com/academy and enter code: 259675

Properties of Operations

The Commutative Property

The commutative property applies to addition and multiplication and states that these operations can be completed in any order. The **commutative property of addition** states that numbers and terms can be added together in any order to still get the same value. For example, $3 + 4 = 7$ and $4 + 3 = 7$. Also, we can use the commutative property of addition to show that $3x + 4 + 2^2$ is equivalent to $4 + 3x + 2^2$ and $2^2 + 4 + 3x$. When adding terms, you can add in any order and get the same value.

The **commutative property of multiplication** states that numbers and terms can be multiplied in any order to get the same value. For example, 12×3 is equivalent to 3×12. Additionally, we can use the commutative property of multiplication to assume that $(5 + 3) \times (36 - 6)$ is equivalent to $(36 - 6) \times (5 + 3)$. You can multiply terms in any order and still get the same value.

The Associative Property

The **associative property of addition** states that if three or more terms are being added together, the value is the same regardless of the groupings.

For example, given the expression $3 + 4 + 6$, these terms can be grouped and added in any form. $3 + 4 + 6$ is equivalent to $(3 + 4) + 6$ and is also equivalent to $3 + (4 + 6)$. This can be applied to write equivalent expressions in a variety of ways.

For example, suppose we are given the expression $5 + (y + 2) + 4$. We can generate equivalent expressions knowing the associative property. Knowing that when three or more terms are added, the grouping is irrelevant, we can say that this expression is equivalent to $5 + y + (2 + 4)$, and it is equivalent to $(5 + y) + (2 + 4)$. It is even equivalent to $5 + y + 2 + 4$.

The **associative property of multiplication** states that if three or more terms are being multiplied together, the value is the same regardless of the grouping. We can use this property to identify and generate equivalent expressions.

For example, given the expression $2 \times 7 \times 3$, these terms can be grouped in any way and still get the same value. $2 \times 7 \times 3$ is equivalent to $(2 \times 7) \times 3$ or $2 \times (7 \times 3)$.

The Identity Property

The **identity property of multiplication** states that when a number is multiplied by 1, you get the same number. That is, anything multiplied by 1 is itself. For example, $2 \times 1 = 2$, or $1 \times -36 = -36$. Using the identity property of multiplication, we can identify and generate equivalent expressions. Let's say that we are given the expression $15 - (3 \times 4)$. We can generate equivalent expressions using the identity property. One equivalent expression example would be $(15 \times 1) - (3 \times 4)$. Another example would be $15 - (1 \times 3 \times 4)$. We can say these expressions are equivalent because the identity property of multiplication states that we can multiply any portion of an expression by 1 to get the same value.

The **identity property of addition** states that when 0 is added to a number, you get the same number. For example, $2 + 0 = 2$, or $0 + -3 = -3$. We can also use this property to identify and generate equivalent expressions. For example, if we are given the expression $2 \times (1 + 2)$, we could write the equivalent expressions $2 \times (0 + 1 + 2)$ or $(2 + 0) \times (1 + 2)$.

The Inverse Property

The **inverse property of addition** states that the sum of a number and its opposite is always equal to 0. Remember, the opposite of a number is a number that is opposite on the number line from zero, or the same number with the opposite sign. For example, −4 is opposite to 4, and 1,726.9 is opposite to −1,726.9. So, the inverse property of addition states that if you add opposite numbers, their sum is zero. For example, $5 + (-5) = 0$ and $-5 + 5 = 0$.

The **inverse property of multiplication** states that a number multiplied by its reciprocal is always equal to 1. The **reciprocal** of a number is its "flipped" fraction. For example, the reciprocal of 5 is $\frac{1}{5}$, or the reciprocal of $\frac{2}{3}$ is $\frac{3}{2}$. The inverse property of multiplication can be applied for these values, $5 \times \frac{1}{5} = 1$ and $\frac{2}{3} \times \frac{3}{2} = 1$. This is because when you multiply across, you get a fraction that is equal to 1.

$$\frac{2}{3} \times \frac{3}{2} = \frac{6}{6} = 1$$

The Distributive Property

The **distributive property** explains how multiplication and addition interact. It says that when multiplying one number by the sum of two other numbers, the same result can also be obtained by multiplying the one number by each of the numbers individually and then adding the products. For example, to multiply 2 by the sum of 7 and 3, the direct approach says, "the sum of 7 and 3 is 10, and 2 times 10 is 20." This would be expressed as $2 \times (7 + 3) = 2 \times 10 = 20$. On the other hand, the distributive property states that the same answer can be achieved by multiplying each number inside the parentheses and adding the products. That is, "the product of 2 and 7 is 14, the product of

2 and 3 is 6, and the sum of 14 and 6 is 20." This would be expressed as $2 \times (7 + 3) = 2 \times 7 + 2 \times 3 = 14 + 6 = 20$, and it is demonstrated below.

$$2 \times (7 + 3) = 2 \times 7 + 2 \times 3$$

This same concept can be used when multiplying a number by the difference of two numbers. For example, $5 \times (10 - 4) = 5 \times 10 - 5 \times 4$. Since $5 \times 10 = 50$ and $5 \times 4 = 20$, the result is $50 - 20 = 30$. This answer can be checked by subtracting inside the parentheses first and then multiplying: $5 \times (10 - 4) = 5 \times 6 = 30$.

Review Video: Commutative, Associative, and Distributive Properties
Visit mometrix.com/academy and enter code: 483176

Properties of Exponents

The properties of exponents are as follows:

Property	Description
$a^1 = a$	Any number to the power of 1 is equal to itself
$1^n = 1$	The number 1 raised to any power is equal to 1
$a^0 = 1$	Any number raised to the power of 0 is equal to 1
$a^n \times a^m = a^{n+m}$	Add exponents to multiply powers of the same base number
$a^n \div a^m = a^{n-m}$	Subtract exponents to divide powers of the same base number
$(a^n)^m = a^{n \times m}$	When a power is raised to a power, the exponents are multiplied
$(a \times b)^n = a^n \times b^n$ $(a \div b)^n = a^n \div b^n$	Multiplication and division operations inside parentheses can be raised to a power. This is the same as each term being raised to that power.
$a^{-n} = \frac{1}{a^n}$	A negative exponent is the same as the reciprocal of a positive exponent

Note that exponents do not have to be integers. Fractional or decimal exponents follow all the rules above as well. Example: $5^{\frac{1}{4}} \times 5^{\frac{3}{4}} = 5^{\frac{1}{4}+\frac{3}{4}} = 5^1 = 5$.

Review Video: Properties of Exponents
Visit mometrix.com/academy and enter code: 532558

Factors and Multiples

Factors and Greatest Common Factor

A whole number a is a **factor** (or **divisor**) of a whole number b if a divides b evenly. In other words, a is a factor of b if the quotient $b \div a$ is a whole number with a remainder of 0. For instance, 3 is a factor of 12 because $12 \div 3 = 4$ with no remainder. Another way to say this is that a is a factor of b if we can multiply a by another whole number to get b. So, we can also show that 3 is a factor of 12 by noting that $3 \times 4 = 12$.

Every positive whole number has 1 and itself as factors. If a whole number greater than one has *only* 1 and itself as factors, we call it a **prime number**. For instance, 5 is a prime number because its only factors are 1 and 5. The first several prime numbers are 2, 3, 5, 7, 11, and 13.

If a whole number greater than 1 is not prime—that is, if it has factors besides 1 and itself—then it is a **composite number.** For instance, 10 is a composite number because it has factors 2 and 5 in addition to 1 and 10. The first several composite numbers are 4, 6, 8, 9, 10, 12, 14, and 15.

A **prime factor** of a whole number is a factor that is also a prime number. For example, the prime factors of 12 are 2 and 3. The prime factors of 15 are 3 and 5.

A **common factor** of two (or more) whole numbers is a number that is a factor of both (or all) of them. For example, the factors of 12 are 1, 2, 3, 4, 6, and 12, while the factors of 15 are 1, 3, 5, and 15. The common factors (underlined) of 12 and 15 are 1 and 3.

The **greatest common factor** (**GCF**) of two (or more) whole numbers is the largest number that is a factor of both (or all) of them. For example, the factors of 15 are 1, 3, 5, and 15; the factors of 35 are 1, 5, 7, and 35. Therefore, the greatest common factor of 15 and 35 is 5.

Review Video: Factors
Visit mometrix.com/academy and enter code: 920086

Review Video: Prime Numbers and Factorization
Visit mometrix.com/academy and enter code: 760669

Multiples and Least Common Multiple

A whole number b is a **multiple** of a whole number a when a is a factor of b. This means that b is the product of a and another whole number. For example, the multiples of 7 are $0 \times 7 = 0, 1 \times 7 = 7, 2 \times 7 = 14, 3 \times 7 = 21, 4 \times 7 = 28, 5 \times 7 = 35$, ... Dividing 0, 7, 14, 21, 28, and 35 by 7 results in the whole numbers 0, 1, 2, 3, 4, and 5, respectively, showing that 7 is a factor of these numbers.

The least common multiple (**LCM**) of two (or more) whole numbers is the smallest number that is a multiple of both (or all) of them. For example, the multiples of 3 are 3, 6, 9, 12, 15, ...; the multiples of 5 are 5, 10, 15, 20, ... The smallest number that appears in both lists is 15, so the least common multiple of 3 and 5 is 15.

Review Video: Multiples
Visit mometrix.com/academy and enter code: 626738

Review Video: Greatest Common Factor and Least Common Multiple
Visit mometrix.com/academy and enter code: 838699

Fractions

A **fraction** is a number that is expressed as one integer written above another integer, with a dividing line between them $\left(\frac{x}{y}\right)$. It represents the **quotient** of the two numbers "x divided by y." It can also be thought of as x out of y equal parts.

The top number of a fraction is called the **numerator**, and it represents the number of parts under consideration. The 1 in $\frac{1}{4}$ means that 1 part out of the whole is being considered in the calculation.

The bottom number of a fraction is called the **denominator**, and it represents the total number of equal parts. The 4 in $\frac{1}{4}$ means that the whole consists of 4 equal parts. A fraction cannot have a denominator of zero; this is referred to as "*undefined.*"

Fractions can be manipulated, without changing the value of the fraction, by multiplying or dividing (but not adding or subtracting) both the numerator and denominator by the same number. If you divide both numbers by a common factor, you are **reducing** or simplifying the fraction. Two fractions that have the same value but are expressed differently are known as **equivalent fractions**. For example, $\frac{2}{10}, \frac{3}{15}, \frac{4}{20}$, and $\frac{5}{25}$ are all equivalent fractions. They can also all be reduced or simplified to $\frac{1}{5}$.

When two fractions are manipulated so that they have the same denominator, this is known as finding a **common denominator**. The number chosen to be that common denominator should be the least common multiple of the two original denominators. Example: $\frac{3}{4}$ and $\frac{5}{6}$; the least common multiple of 4 and 6 is 12. Manipulating to achieve the common denominator: $\frac{3}{4} = \frac{9}{12}; \frac{5}{6} = \frac{10}{12}$.

Review Video: Overview of Fractions
Visit mometrix.com/academy and enter code: 262335

Proper Fractions and Mixed Numbers

A fraction whose denominator is greater than its numerator is known as a **proper fraction**, while a fraction whose numerator is greater than its denominator is known as an **improper fraction**. Proper fractions have values *less than one* and improper fractions have values *greater than one.*

A **mixed number** is a number that contains both an integer and a fraction. Any improper fraction can be rewritten as a mixed number. Example: $\frac{8}{3} = \frac{6}{3} + \frac{2}{3} = 2 + \frac{2}{3} = 2\frac{2}{3}$. Similarly, any mixed number can be rewritten as an improper fraction. Example: $1\frac{3}{5} = 1 + \frac{3}{5} = \frac{5}{5} + \frac{3}{5} = \frac{8}{5}$.

Review Video: Proper and Improper Fractions and Mixed Numbers
Visit mometrix.com/academy and enter code: 211077

Adding and Subtracting Fractions

If two fractions have a common denominator, they can be added or subtracted simply by adding or subtracting the two numerators and retaining the same denominator. If the two fractions do not already have the same denominator, one or both of them must be manipulated to achieve a common denominator before they can be added or subtracted. Example: $\frac{1}{2} + \frac{1}{4} = \frac{2}{4} + \frac{1}{4} = \frac{3}{4}$.

Review Video: Adding and Subtracting Fractions
Visit mometrix.com/academy and enter code: 378080

Multiplying Fractions

Two fractions can be multiplied by multiplying the two numerators to find the new numerator and the two denominators to find the new denominator. Example: $\frac{1}{3} \times \frac{2}{3} = \frac{1\times2}{3\times3} = \frac{2}{9}$.

Dividing Fractions

Two fractions can be divided by flipping the numerator and denominator of the second fraction and then proceeding as though it were a multiplication problem. Example: $\frac{2}{3} \div \frac{3}{4} = \frac{2}{3} \times \frac{4}{3} = \frac{8}{9}$.

Review Video: Multiplying and Dividing Fractions
Visit mometrix.com/academy and enter code: 473632

Multiplying a Mixed Number by a Whole Number or a Decimal

When multiplying a mixed number by something, it is usually best to convert it to an improper fraction first. Additionally, if the multiplicand is a decimal, it is most often simplest to convert it to a fraction. For instance, to multiply $4\frac{3}{8}$ by 3.5, begin by rewriting each quantity as a whole number plus a proper fraction. Remember, a mixed number is a fraction added to a whole number and a decimal is a representation of the sum of fractions, specifically tenths, hundredths, thousandths, and so on:

$$4\frac{3}{8} \times 3.5 = \left(4 + \frac{3}{8}\right) \times \left(3 + \frac{1}{2}\right)$$

Next, the quantities being added need to be expressed with the same denominator. This is achieved by multiplying and dividing the whole number by the denominator of the fraction. Recall that a whole number is equivalent to that number divided by 1:

$$= \left(\frac{4}{1} \times \frac{8}{8} + \frac{3}{8}\right) \times \left(\frac{3}{1} \times \frac{2}{2} + \frac{1}{2}\right)$$

When multiplying fractions, remember to multiply the numerators and denominators separately:

$$= \left(\frac{4 \times 8}{1 \times 8} + \frac{3}{8}\right) \times \left(\frac{3 \times 2}{1 \times 2} + \frac{1}{2}\right)$$
$$= \left(\frac{32}{8} + \frac{3}{8}\right) \times \left(\frac{6}{2} + \frac{1}{2}\right)$$

Now that the fractions have the same denominators, they can be added:

$$= \frac{35}{8} \times \frac{7}{2}$$

Finally, perform the last multiplication and then simplify:

$$= \frac{35 \times 7}{8 \times 2} = \frac{245}{16} = \frac{240}{16} + \frac{5}{16} = 15\frac{5}{16}$$

Comparing Fractions

It is important to master the ability to compare and order fractions. This skill is relevant to many real-world scenarios. For example, carpenters often compare fractional construction nail lengths when preparing for a project, and bakers often compare fractional measurements to have the correct ratio of ingredients. There are three commonly used strategies when comparing fractions. These strategies are referred to as the common denominator approach, the decimal approach, and the cross-multiplication approach.

USING A COMMON DENOMINATOR TO COMPARE FRACTIONS

The fractions $\frac{2}{3}$ and $\frac{4}{7}$ have different denominators. $\frac{2}{3}$ has a denominator of 3, and $\frac{4}{7}$ has a denominator of 7. In order to precisely compare these two fractions, it is necessary to use a common denominator. A common denominator is a common multiple that is shared by both denominators. In this case, the denominators 3 and 7 share a multiple of 21. In general, it is most efficient to select the least common multiple for the two denominators.

Rewrite each fraction with the common denominator of 21. Then, calculate the new numerators as illustrated below.

$$\frac{2}{3} = \frac{\mathbf{14}}{21} \quad (\times 7) \qquad \frac{4}{7} = \frac{\mathbf{12}}{21} \quad (\times 3)$$

For $\frac{2}{3}$, multiply the numerator and denominator by 7. The result is $\frac{14}{21}$.

For $\frac{4}{7}$, multiply the numerator and denominator by 3. The result is $\frac{12}{21}$.

Now that both fractions have a denominator of 21, the fractions can accurately be compared by comparing the numerators. Since 14 is greater than 12, the fraction $\frac{14}{21}$ is greater than $\frac{12}{21}$. This means that $\frac{2}{3}$ is greater than $\frac{4}{7}$.

USING DECIMALS TO COMPARE FRACTIONS

Sometimes decimal values are easier to compare than fraction values. For example, $\frac{5}{8}$ is equivalent to 0.625 and $\frac{3}{5}$ is equivalent to 0.6. This means that the comparison of $\frac{5}{8}$ and $\frac{3}{5}$ can be determined by comparing the decimals 0.625 and 0.6. When both decimal values are extended to the thousandths place, they become 0.625 and 0.600, respectively. It becomes clear that 0.625 is greater than 0.600 because 625 thousandths is greater than 600 thousandths. In other words, $\frac{5}{8}$ is greater than $\frac{3}{5}$ because 0.625 is greater than 0.6.

USING CROSS-MULTIPLICATION TO COMPARE FRACTIONS

Cross-multiplication is an efficient strategy for comparing fractions. This is a shortcut for the common denominator strategy. Start by writing each fraction next to one another. Multiply the numerator of the fraction on the left by the denominator of the fraction on the right. Write down the result next to the fraction on the left. Now multiply the numerator of the fraction on the right by the denominator of the fraction on the left. Write down the result next to the fraction on the right. Compare both products. The fraction with the larger result is the larger fraction.

Consider the fractions $\frac{4}{7}$ and $\frac{5}{9}$.

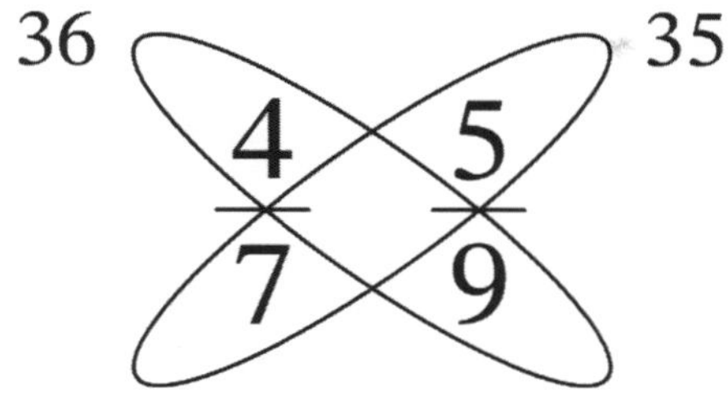

36 is greater than 35. Therefore, $\frac{4}{7}$ is greater than $\frac{5}{9}$.

Decimals

Decimals are one way to represent parts of a whole. Using the place value system, each digit to the right of a decimal point denotes the number of units of a corresponding *negative* power of ten. For example, consider the decimal 0.24. We can use a model to represent the decimal. Since a dime is worth one-tenth of a dollar and a penny is worth one-hundredth of a dollar, one possible model to represent this fraction is to have 2 dimes representing the 2 in the tenths place and 4 pennies representing the 4 in the hundredths place:

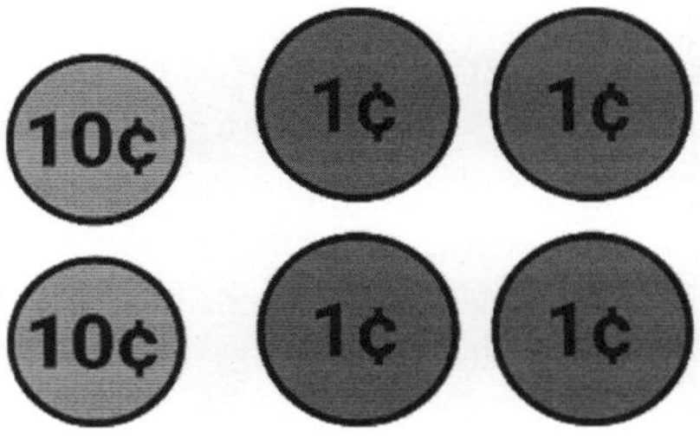

To write the decimal as a fraction, put the decimal in the numerator with 1 in the denominator. Multiply the numerator and denominator by tens until there are no more decimal places. Then simplify the fraction to lowest terms. For example, converting 0.24 to a fraction:

$$0.24 = \frac{0.24}{1} = \frac{0.24 \times 100}{1 \times 100} = \frac{24}{100} = \frac{6}{25}$$

Review Video: Decimals
Visit mometrix.com/academy and enter code: 837268

Operations with Decimals

Adding and Subtracting Decimals

When adding and subtracting decimals, the decimal points must always be aligned. Adding decimals is just like adding regular whole numbers. Example: $4.5 + 2.0 = 6.5$.

If the problem-solver does not properly align the decimal points, an incorrect answer of 4.7 may result. An easy way to add decimals is to align all of the decimal points in a vertical column visually. This will allow you to see exactly where the decimal should be placed in the final answer. Begin adding from right to left. Add each column in turn, making sure to carry the number to the left if a column adds up to more than 9. The same rules apply to the subtraction of decimals.

Review Video: Adding and Subtracting Decimals
Visit mometrix.com/academy and enter code: 381101

Multiplying Decimals

A simple multiplication problem has two components: a **multiplicand** and a **multiplier**. When multiplying decimals, work as though the numbers were whole rather than decimals. Once the final product is calculated, count the number of places to the right of the decimal in both the multiplicand and the multiplier. Then, count that number of places from the right of the product and place the decimal in that position.

For example, 12.3×2.56 has a total of three places to the right of the respective decimals. Multiply 123×256 to get 31,488. Now, beginning on the right, count three places to the left and insert the decimal. The final product will be 31.488.

Review Video: How to Multiply Decimals
Visit mometrix.com/academy and enter code: 731574

Dividing Decimals

Every division problem has a **divisor** and a **dividend**. The dividend is the number that is being divided. In the problem $14 \div 7$, 14 is the dividend and 7 is the divisor. In a division problem with decimals, the divisor must be converted into a whole number. Begin by moving the decimal in the divisor to the right until a whole number is created. Next, move the decimal in the dividend the same number of spaces to the right. For example, 4.9 into 24.5 would become 49 into 245. The decimal was moved one space to the right to create a whole number in the divisor, and then the same was done for the dividend. Once the whole numbers are created, the problem is carried out normally: $245 \div 49 = 5$.

Review Video: Dividing Decimals
Visit mometrix.com/academy and enter code: 560690

Review Video: Dividing Decimals by Whole Numbers
Visit mometrix.com/academy and enter code: 535669

Percentages

Percentages can be thought of as fractions that are based on a whole of 100; that is, one whole is equal to 100%. The word **percent** means "per hundred." Percentage problems are often presented in three main ways:

- Find what percentage of some number another number is.
    - Example: What percentage of 40 is 8?
- Find what number is some percentage of a given number.
    - Example: What number is 20% of 40?
- Find what number another number is a given percentage of.
    - Example: What number is 8 20% of?

There are three components in each of these cases: a **whole** (W), a **part** (P), and a **percentage** (%). These are related by the equation: $P = W \times \%$. This can easily be rearranged into other forms that may suit different questions better: $\% = \frac{P}{W}$ and $W = \frac{P}{\%}$. Percentage problems are often also word problems. As such, a large part of solving them is figuring out which quantities are what. For example, consider the following word problem:

In a school cafeteria, 7 students choose pizza, 9 choose hamburgers, and 4 choose tacos. What percentage of students choose tacos?

To find the whole, you must first add all of the parts: $7 + 9 + 4 = 20$. The percentage can then be found by dividing the part by the whole $\left(\% = \frac{P}{W}\right)$: $\frac{4}{20} = \frac{20}{100} = 20\%$.

Review Video: Computation with Percentages
Visit mometrix.com/academy and enter code: 693099

Calculating Percent Change

Suppose a quantity has a particular value (the *old value*) and then we add something (the *change*) to it to get another value (the *new value*). We can describe this process by the simple equation (old value) + change = (new value). If we know the old and new values, we can rearrange this equation to find the change, getting change = (new value) − (old value). For instance, if a store's price for a box of computer paper goes from \$20 last week to \$25 this week, this is a change of (new value) − (old value) = \$25 − \$20 = \$5. Or, if the size of the freshman class at a college goes from 500 students one year to 440 students the next year, this is a change of (new value) − (old value) = 440 − 500 = −60 students. So, we see that change can be positive or negative.

Instead of the word *change*, we sometimes use the words *increase* or *decrease* to specify whether the value goes up or down, respectively. In the examples above, the price of computer paper increases by \$5 and the freshman class decreases by 60 students. Note that the decrease is 60 students and not −60 because the word *decrease* already means that the value goes down. So, *increase* is the same as positive change and *decrease* is the opposite or negative change.

If the changing quantity represents an amount (how much of something there is), we can also calculate the **percent change**. This is the change expressed as a percentage of the old amount. To calculate this, we divide the change by the old amount and express the quotient as a percent. That is, we use the formula percent change $= \frac{\text{change}}{\text{old value}}$, converting the resulting decimal answer to a percent. In the examples above, the price of a box of computer paper has a percent change of $\frac{\text{change in price}}{\text{old price}} = \frac{\$5}{\$20} = 0.25 = 25\%$, and the size of the freshman class at the college has a percent change of $\frac{\text{change in enrollment}}{\text{old enrollment}} = \frac{-60}{500} = -0.12 = -12\%$. We can also use the terms *percent increase* and *percent decrease*, saying that the price of computer paper increases by 25% and the size of the freshman class decreases by 12%. Note that the denominator is always the old amount, never the new amount.

Example: Your landlord raises your rent from \$1,500 to \$1,700 per month. To find the percent change in your rent (rounded to the nearest tenth of a percent), you calculate as follows.

$$\begin{aligned}\text{percent change in rent} &= \frac{\text{change in rent}}{\text{old rent}} = \frac{(\text{new rent}) - (\text{old rent})}{\text{old rent}} \\ &= \frac{\$1{,}700 - \$1{,}500}{\$1{,}500} = \frac{\$200}{\$1{,}500} = 0.1333 \ldots \approx 13.3\%\end{aligned}$$

Therefore, the percent change in your rent is approximately 13.3%.

Review Video: Percent Change
Visit mometrix.com/academy and enter code: 907890

Converting Between Percentages, Fractions, and Decimals

Converting decimals to percentages and percentages to decimals is as simple as moving the decimal point. To *convert from a decimal to a percentage*, move the decimal point **two places to the right**. To *convert from a percentage to a decimal*, move it **two places to the left**. It may be helpful to remember that the percentage number will always be larger than the equivalent decimal number. Example:

$$\begin{array}{ccc} 0.23 = 23\% & 5.34 = 534\% & 0.007 = 0.7\% \\ 700\% = 7.00 & 86\% = 0.86 & 0.15\% = 0.0015 \end{array}$$

To convert a fraction to a decimal, simply divide the numerator by the denominator in the fraction. To convert a decimal to a fraction, put the decimal in the numerator with 1 in the denominator. Multiply the numerator and denominator by tens until there are no more decimal places. Then simplify the fraction to lowest terms. For example, converting 0.24 to a fraction:

$$0.24 = \frac{0.24}{1} = \frac{0.24 \times 100}{1 \times 100} = \frac{24}{100} = \frac{6}{25}$$

Fractions can be converted to a percentage by finding equivalent fractions with a denominator of 100. Example:

$$\frac{7}{10} = \frac{70}{100} = 70\% \quad \frac{1}{4} = \frac{25}{100} = 25\%$$

To convert a percentage to a fraction, divide the percentage number by 100 and reduce the fraction to its simplest possible terms. Example:

$$60\% = \frac{60}{100} = \frac{3}{5} \quad 96\% = \frac{96}{100} = \frac{24}{25}$$

Review Video: Converting Fractions to Percentages and Decimals
Visit mometrix.com/academy and enter code: 306233

Review Video: Converting Percentages to Decimals and Fractions
Visit mometrix.com/academy and enter code: 287297

Review Video: Converting Decimals to Fractions and Percentages
Visit mometrix.com/academy and enter code: 986765

Review Video: Converting Decimals, Improper Fractions, and Mixed Numbers
Visit mometrix.com/academy and enter code: 696924

Rational and Irrational Numbers

The term **rational** means that the number can be expressed as a ratio or fraction. That is, a number, r, is rational if and only if it can be represented by a fraction $\frac{a}{b}$ where a and b are integers and b does not equal 0. The set of rational numbers includes integers and decimals. If there is no finite way to represent a value with a fraction of integers, then the number is **irrational**. Common irrational numbers are π and the square roots of whole numbers that are not perfect squares (e.g., $\sqrt{5}$ or $\sqrt{21}$). The sum or product of an integer and an irrational number is always irrational (e.g., 3π or $7 + \sqrt{6}$).

Review Video: Rational and Irrational Numbers
Visit mometrix.com/academy and enter code: 280645

Review Video: Ordering Rational Numbers
Visit mometrix.com/academy and enter code: 419578

Review Video: Irrational Numbers on a Number Line
Visit mometrix.com/academy and enter code: 433866

Proportions and Ratios

Proportions

There is a **proportion** between two variable quantities if there is a constant relationship between their products or quotients, a relationship that does not change as the quantities themselves change.

Given variable quantities x and y, we say that they are **directly proportional** (or that y **varies directly with** x) if their quotient or *ratio* is constant—that is, if there is a constant k such that $\frac{y}{x} = k$ is always true. Another way of saying this is that y is a constant multiple of x, so that $y = kx$ is always true. We call the number k the **constant of proportionality**. For example, if you drive at a constant 50 miles per hour, then the distance, y, that you travel in miles is 50 times the number of hours, x, that you drive. In symbols, $y = 50x$ miles (or $\frac{y}{x} = 50$ mph). So, the distance you travel, y, is directly proportional to (or varies directly with) the time you travel, x, with constant of proportionality $k = 50$ mph.

The quantities x and y are **inversely proportional** (or y varies inversely with x) if their product is constant—that is, if there is a constant k such that $xy = k$ is always true. Another way of saying this is to say that y is a constant multiple of the reciprocal of x so that $y = \frac{k}{x}$ is always true. For instance, suppose you drive at speed (rate) y mph for x hours, going a total of 120 miles. Since rate × time = distance, we get $xy = 120$ miles (or $y = \frac{120}{x}$ miles per hour). Thus, your driving speed, y, is inversely proportional to (or varies inversely with) your drive time, x, with constant of proportionality $k = 120$ miles.

Review Video: Proportions
Visit mometrix.com/academy and enter code: 505355

Ratios

A **ratio** expresses the sizes of two quantities relative to each other. For instance, suppose we have 3 copies of sheet music to share among 6 singers. We can divide the singers into groups of 2 and give each group 1 copy of the music. Thus, there is 1 copy of the music for every 2 singers, and we say that the **ratio** of sheet music to singers is 1 to 2, which we write either as a fraction $\frac{1}{2}$ or using a colon $1 : 2$. Of course, it is also true there are 3 copies for every 6 singers so that the ratio of sheet music to singers is also 3 to 6, which we write as $\frac{3}{6}$ or $3 : 6$. So, the ratios $\frac{1}{2}$ and $\frac{3}{6}$ express the same relative quantities of music and singers. We say that these ratios are equal or **equivalent**, and we note that ratios are equal precisely when their fractions are equal (so, in this case, $\frac{1}{2} = \frac{3}{6}$ as fractions). We can also express the quantities in the other order and say that the ratio of singers to music is $\frac{2}{1}$ or $2 : 1$ (or $\frac{6}{3}$ or $6 : 3$).

Review Video: Ratios
Visit mometrix.com/academy and enter code: 996914

Constant of Proportionality

If variable quantities x and y are proportional and we know a pair of corresponding values for them, then we can find their constant of proportionality. If they are directly proportional, we use the formula $\frac{y}{x} = k$. If they are inversely proportional, we use the formula $xy = k$.

Example: The cost in dollars, y, of buying fence posts is directly proportional to the number, x, that you buy. If it costs \$51 to buy 17 fence posts, what is the constant of proportionality? Because of direct proportionality, we know that $\frac{y}{x} = k$. Since this works for every pair of corresponding x- and y-values, it also works for $x = 17$ and $y = 51$. This gives us $\frac{51}{17} = k$, which simplifies to $k = 3$. Note also that this is the unit price, namely \$3 per fence post.

Work/Unit Rate

Unit rate expresses a quantity of one thing in terms of one unit of another. For example, if you travel 30 miles every two hours, a unit rate expresses this comparison in terms of one hour: in one hour you travel 15 miles, so your unit rate is 15 miles per hour. Other examples are how much one ounce of food costs (price per ounce) or figuring out how much one egg costs out of the dozen (price per 1 egg, instead of price per 12 eggs). The denominator of a unit rate is always 1. Unit rates are used to compare different situations to solve problems. For example, to make sure you get the best deal when deciding which kind of soda to buy, you can find the unit rate of each. If soda #1 costs $1.50 for a 1-liter bottle, and soda #2 costs $2.75 for a 2-liter bottle, it would be a better deal to buy soda #2, because its unit rate is only $1.375 per liter, which is cheaper than soda #1. Unit rates can also help determine the length of time a given event will take. For example, if you can paint 2 rooms in 4.5 hours, you can determine how long it will take you to paint 5 rooms by solving for the unit rate per room and then multiplying that by 5.

> **Review Video: Rates and Unit Rates**
> Visit mometrix.com/academy and enter code: 185363

Cross Multiplication

Finding an Unknown in Equivalent Expressions

It is often necessary to apply information given about a rate or proportion to a new scenario. For example, if you know that Jedha can run a marathon (26.2 miles) in 3 hours, how long would it take her to run 10 miles at the same pace? Start by setting up equivalent expressions:

$$\frac{26.2\text{ mi}}{3\text{ hr}} = \frac{10\text{ mi}}{x\text{ hr}}$$

Now, cross multiply and solve for x:

$$\begin{aligned} 26.2x &= 30 \\ x &= \frac{30}{26.2} = \frac{15}{13.1} \\ x &\approx 1.15\text{ hrs } \textit{or } 1\text{ hr } 9\text{ min} \end{aligned}$$

So, at this pace, Jedha could run 10 miles in about 1.15 hours or about 1 hour and 9 minutes.

> **Review Video: Cross Multiplying Fractions**
> Visit mometrix.com/academy and enter code: 893904

Slope

Finding Slope Given Graph or Table

On a graph with two points, (x_1, y_1) and (x_2, y_2), the **slope** is found with the formula $m = \frac{y_2 - y_1}{x_2 - x_1}$; where $x_1 \neq x_2$ and m stands for slope. If the value of the slope is **positive**, the line has an *upward direction* from left to right. If the value of the slope is **negative**, the line has a *downward direction* from left to right. Consider the following example:

A new book goes on sale in bookstores and online stores. In the first month, 5,000 copies of the book are sold. Over time, the book continues to grow in popularity. The data for the number of copies sold is in the table below.

# of Months on Sale	1	2	3	4	5
# of Copies Sold (In Thousands)	5	10	15	20	25

So, the number of copies that are sold and the time that the book is on sale is a proportional relationship. In this example, an equation can be used to show the data: $y = 5x$, where x is the number of months that the book is on sale. Also, y is the number of copies sold. So, the slope of the corresponding line is $\frac{\text{rise}}{\text{run}} = \frac{5}{1} = 5$.

Finding Slope Given an Equation

When given an equation of a line, it is necessary to solve for y to determine the slope of the line. Given the equation $6x + 2y = 8$, find the slope. First, subtract $6x$ from both sides of the equation, resulting in $2y = -6x + 8$. Then divide both sides of the equation by 2, resulting in $y = -3x + 4$. This then allows us to conclude that the slope of the line is $m = -3$, the coefficient of x. Once an equation is in the form $y = mx + b$, the slope and y-intercept can easily be determined. For this reason, we refer to the equation $y = mx + b$ as "slope-intercept form" of the equation of a line.

Review Video: Finding the Slope of a Line
Visit mometrix.com/academy and enter code: 766664

Complex Numbers

Complex numbers consist of a real component and an imaginary component. Complex numbers are expressed in the form $a + bi$ with real component a and imaginary component bi. The imaginary unit i is equal to $\sqrt{-1}$. That means $i^2 = -1$. The imaginary unit provides a way to find the square root of a negative number. For example, $\sqrt{-25}$ is $5i$. You should expect questions asking you to add, subtract, multiply, divide, and simplify complex numbers. You may see a question that says, "Add $3 + 2i$ and $5 - 7i$" or "Subtract $4 + i\sqrt{5}$ from $2 + i\sqrt{5}$." Or you may see a question that says, "Multiply $6 + 2i$ by $8 - 4i$" or "Divide $1 - 3i$ by $9 - 7i$."

Operations on Complex Numbers

Operations with complex numbers resemble operations with variables in algebra. When adding or subtracting complex numbers, you can only combine like terms—real terms with real terms and imaginary terms with imaginary terms. For example, if you are asked to simplify the expression $-2 + 4i - (-3 + 7i) - 5i$, you should first remove the parentheses to yield $-2 + 4i + 3 - 7i - 5i$. Combining like terms yields $1 - 8i$. One interesting aspect of imaginary numbers is that if i has an exponent greater than 1, it can be simplified. Example: $i^2 = -1$, $i^3 = -i$, and $i^4 = 1$. When multiplying complex numbers, remember to simplify each i with an exponent greater than 1. For example, you might see a question that says, "Simplify $(2 - i)(3 + 2i)$." You need to distribute and multiply to get $6 + 4i - 3i - 2i^2$. This is further simplified to $6 + i - 2(-1)$, or $8 + i$.

Simplifying Expressions with Complex Denominators

If an expression contains an i in the denominator, it must be simplified. Remember, roots cannot be left in the denominator of a fraction. Since i is equivalent to $\sqrt{-1}$, i cannot be left in the denominator of a fraction. You must rationalize the denominator of a fraction that contains a complex denominator by multiplying the numerator and denominator by the conjugate of the denominator. The conjugate of the complex number $a + bi$ is $a - bi$. You can simplify $\frac{2}{5i}$ by simply multiplying $\frac{2}{5i} \times \frac{i}{i}$, which yields $-\frac{2}{5}i$. And you can simplify $\frac{5+3i}{2-4i}$ by multiplying $\frac{5+3i}{2-4i} \times \frac{2+4i}{2+4i}$. This yields $\frac{10+20i+6i-12}{4-8i+8i+16}$ which simplifies to $\frac{-2+26i}{20}$ or $\frac{-1+13i}{10}$, which can also be written as $-\frac{1}{10} + \frac{13}{10}i$.

Matrix Basics

A **matrix** (plural: matrices) is a rectangular array of numbers or variables, often called **elements**, which are arranged in columns and rows. A matrix is generally represented by a capital letter, with its elements represented by the corresponding lowercase letter with two subscripts indicating the row and column of the element. For example, n_{ab} represents the element in row a column b of matrix N.

$$N = \begin{bmatrix} n_{11} & n_{12} & n_{13} \\ n_{21} & n_{22} & n_{23} \end{bmatrix}$$

A matrix can be described in terms of the number of rows and columns it contains in the format $a \times b$, where a is the number of rows and b is the number of columns. The matrix shown above is a 2×3 matrix. Any $a \times b$ matrix where $a = b$ is a square matrix. A **vector** is a matrix that has exactly one column (**column vector**) or exactly one row (**row vector**).

The **main diagonal** of a matrix is the set of elements on the diagonal from the top left to the bottom right of a matrix. Because of the way it is defined, only square matrices will have a main diagonal. For the matrix shown below, the main diagonal consists of the elements $n_{11}, n_{22}, n_{33}, n_{44}$.

$$\begin{bmatrix} n_{11} & n_{12} & n_{13} & n_{14} \\ n_{21} & n_{22} & n_{23} & n_{24} \\ n_{31} & n_{32} & n_{33} & n_{34} \\ n_{41} & n_{42} & n_{43} & n_{44} \end{bmatrix}$$

A 3×4 matrix such as the one shown below would not have a main diagonal because there is no straight line of elements between the top left corner and the bottom right corner that joins the elements.

$$\begin{bmatrix} n_{11} & n_{12} & n_{13} & n_{14} \\ n_{21} & n_{22} & n_{23} & n_{24} \\ n_{31} & n_{32} & n_{33} & n_{34} \end{bmatrix}$$

A **diagonal matrix** is a square matrix that has a zero for every element in the matrix except the elements on the main diagonal. All the elements on the main diagonal must be nonzero numbers.

$$\begin{bmatrix} n_{11} & 0 & 0 & 0 \\ 0 & n_{22} & 0 & 0 \\ 0 & 0 & n_{33} & 0 \\ 0 & 0 & 0 & n_{44} \end{bmatrix}$$

If every element on the main diagonal of a diagonal matrix is equal to one, the matrix is called an **identity matrix**. The identity matrix is often represented by the letter I.

$$I = \begin{bmatrix} 1 & 0 & 0 & 0 \\ 0 & 1 & 0 & 0 \\ 0 & 0 & 1 & 0 \\ 0 & 0 & 0 & 1 \end{bmatrix}$$

A **zero matrix** is a matrix that has zero as the value for every element in the matrix.

$$\begin{bmatrix} 0 & 0 & 0 & 0 \\ 0 & 0 & 0 & 0 \\ 0 & 0 & 0 & 0 \\ 0 & 0 & 0 & 0 \end{bmatrix}$$

The zero matrix is the *identity for matrix addition*. Do not confuse the zero matrix with the identity matrix.

The **negative of a matrix** is also known as the additive inverse of a matrix. If matrix N is the given matrix, then matrix $-N$ is its negative. This means that every element n_{ab} is equal to $-n_{ab}$ in the negative. To find the negative of a given matrix, change the sign of every element in the matrix and keep all elements in their original corresponding positions in the matrix.

If two matrices have the same order and all corresponding elements in the two matrices are the same, then the two matrices are **equal matrices**.

A matrix N may be **transposed** to matrix N^T by changing all rows into columns and changing all columns into rows. The easiest way to accomplish this is to swap the positions of the row and column notations for each element. For example, suppose the element in the second row of the third column of matrix N is $n_{23} = 6$. In the transposed matrix N^T, the transposed element would be $n_{32} = 6$, and it would be placed in the third row of the second column.

$$N = \begin{bmatrix} 1 & 2 & 3 \\ 4 & 5 & 6 \end{bmatrix}; \; N^T = \begin{bmatrix} 1 & 4 \\ 2 & 5 \\ 3 & 6 \end{bmatrix}$$

To quickly transpose a matrix by hand, begin with the first column and rewrite a new matrix with those same elements in the same order in the first row. Write the elements from the second column of the original matrix in the second row of the transposed matrix. Continue this process until all columns have been completed. If the original matrix is identical to the transposed matrix, the matrices are symmetric.

Basic Operations with Matrices

There are two categories of basic operations with regard to matrices: operations between a matrix and a scalar, and operations between two matrices.

Scalar Operations

A scalar being added to a matrix is treated as though it were being added to each element of the matrix. The same is true for subtraction, multiplication and division:

$$M + \mathrm{k} = \begin{bmatrix} m_{11} + k & m_{12} + k \\ m_{21} + k & m_{22} + k \end{bmatrix}$$

$$M - k = \begin{bmatrix} m_{11} - k & m_{12} - k \\ m_{21} - k & m_{22} - k \end{bmatrix}$$

$$M \times k = \begin{bmatrix} m_{11} \times k & m_{12} \times k \\ m_{21} \times k & m_{22} \times k \end{bmatrix}$$

$$M \div k = \begin{bmatrix} m_{11} \div k & m_{12} \div k \\ m_{21} \div k & m_{22} \div k \end{bmatrix}$$

Matrix Addition and Subtraction

All four of the basic operations can be used with operations between matrices (although division is usually discarded in favor of multiplication by the inverse), but there are restrictions on the situations in which they can be used. Matrices that meet all the qualifications for a given operation are called **conformable matrices**. However, conformability is specific to the operation; two matrices that are conformable for addition are not necessarily conformable for multiplication.

For two matrices to be conformable for addition or subtraction, they must be of the same dimension; otherwise, the operation is not defined. If matrix M is a 3×2 matrix and matrix N is a 2×3 matrix, the operations $M + N$ and $M - N$ are meaningless. If matrices M and N are the same size, the operation is as simple as adding or subtracting all of the corresponding elements:

$$\begin{bmatrix} m_{11} & m_{12} \\ m_{21} & m_{22} \end{bmatrix} + \begin{bmatrix} n_{11} & n_{12} \\ n_{21} & n_{22} \end{bmatrix} = \begin{bmatrix} m_{11} + n_{11} & m_{12} + n_{12} \\ m_{21} + n_{21} & m_{22} + n_{22} \end{bmatrix}$$

$$\begin{bmatrix} m_{11} & m_{12} \\ m_{21} & m_{22} \end{bmatrix} - \begin{bmatrix} n_{11} & n_{12} \\ n_{21} & n_{22} \end{bmatrix} = \begin{bmatrix} m_{11} - n_{11} & m_{12} - n_{12} \\ m_{21} - n_{21} & m_{22} - n_{22} \end{bmatrix}$$

The result of addition or subtraction is a matrix of the same dimension as the two original matrices involved in the operation.

Matrix Multiplication

The first thing it is necessary to understand about matrix multiplication is that it is not commutative. In scalar multiplication, the operation is commutative, meaning that $a \times b = b \times a$. For matrix multiplication, this is not the case: $A \times B \neq B \times A$. The terminology must be specific when describing matrix multiplication. The operation $A \times B$ can be described as A multiplied (or **post-multiplied**) by B, or B **pre-multiplied** by A.

For two matrices to be conformable for multiplication, they need not be of the same dimension, but specific dimensions must correspond. Taking the example of two matrices M and N to be multiplied $M \times N$, matrix M must have the same number of columns as matrix N has rows. Put another way, if matrix M has the dimensions $a \times b$ and matrix N has the dimensions $c \times d$, b must equal c if the two matrices are to be conformable for this multiplication. The matrix that results from the multiplication will have the dimensions $a \times d$. If a and d are both equal to 1, the product is simply a scalar. Square matrices of the same dimensions are always conformable for multiplication, and their product is always a matrix of the same size.

The simplest type of matrix multiplication is a 1×2 matrix (a row vector) times a 2×1 matrix (a column vector). These will multiply in the following way:

$$\begin{bmatrix} m_{11} & m_{12} \end{bmatrix} \times \begin{bmatrix} n_{11} \\ n_{21} \end{bmatrix} = m_{11}n_{11} + m_{12}n_{21}$$

The two matrices are conformable for multiplication because matrix M has the same number of columns as matrix N has rows. Because the other dimensions are both 1, the result is a scalar. Expanding our matrices to 1×3 and 3×1, the process is the same:

$$\begin{bmatrix} m_{11} & m_{12} & m_{13} \end{bmatrix} \times \begin{bmatrix} n_{11} \\ n_{21} \\ n_{31} \end{bmatrix} = m_{11}n_{11} + m_{12}n_{21} + m_{13}n_{31}$$

Once again, the result is a scalar. This type of basic matrix multiplication is the building block for the multiplication of larger matrices.

To multiply larger matrices, treat each **row from the first matrix** and each **column from the second matrix** as individual vectors and follow the pattern for multiplying vectors. The scalar value found from multiplying the first-row vector by the first column vector is placed in the first row, first column of the new matrix. The scalar value found from multiplying the second-row vector by the first column vector is placed in the second row, first column of the new matrix. Continue this pattern until each row of the first matrix has been multiplied by each column of the second matrix.

Below is an example of the multiplication of a 3 × 2 matrix and a 2 × 3 matrix.

$$\begin{bmatrix} m_{11} & m_{12} \\ m_{21} & m_{22} \\ m_{31} & m_{32} \end{bmatrix} \times \begin{bmatrix} n_{11} & n_{12} & n_{13} \\ n_{21} & n_{22} & n_{23} \end{bmatrix} = \begin{bmatrix} m_{11}n_{11} + m_{12}n_{21} & m_{11}n_{12} + m_{12}n_{22} & m_{11}n_{13} + m_{12}n_{23} \\ m_{21}n_{11} + m_{22}n_{21} & m_{21}n_{12} + m_{22}n_{22} & m_{21}n_{13} + m_{22}n_{23} \\ m_{31}n_{11} + m_{32}n_{21} & m_{31}n_{12} + m_{32}n_{22} & m_{31}n_{13} + m_{32}n_{23} \end{bmatrix}$$

This process starts by taking the first column of the second matrix and running it through each row of the first matrix. Removing all but the first M row and first N column, we would see only the following:

$$[m_{11} \ \ m_{12}] \times \begin{bmatrix} n_{11} \\ n_{21} \end{bmatrix} = m_{11}n_{11} + m_{12}n_{21}$$

The first product would then be $m_{11}n_{11} + m_{12}n_{21}$. This process will be continued for each column of the N matrix to find the first full row of the product matrix, as shown below.

$$[m_{11}n_{11} + m_{12}n_{21} \quad m_{11}n_{12} + m_{12}n_{22} \quad m_{11}n_{13} + m_{12}n_{23}]$$

After completing the first row, the next step would be to simply move to the second row of the M matrix and repeat the process until all of the rows have been finished. The result is a 3 × 3 matrix.

$$\begin{bmatrix} m_{11} & m_{12} \\ m_{21} & m_{22} \\ m_{31} & m_{32} \end{bmatrix} \times \begin{bmatrix} n_{11} & n_{12} & n_{13} \\ n_{21} & n_{22} & n_{23} \end{bmatrix} = \begin{bmatrix} m_{11}n_{11} + m_{12}n_{21} & m_{11}n_{12} + m_{12}n_{22} & m_{11}n_{13} + m_{12}n_{23} \\ m_{21}n_{11} + m_{22}n_{21} & m_{21}n_{12} + m_{22}n_{22} & m_{21}n_{13} + m_{22}n_{23} \\ m_{31}n_{11} + m_{32}n_{21} & m_{31}n_{12} + m_{32}n_{22} & m_{31}n_{13} + m_{32}n_{23} \end{bmatrix}$$

If the operation were done in reverse ($N \times M$), the result would be a 2 × 2 matrix.

$$\begin{bmatrix} n_{11} & n_{12} & n_{13} \\ n_{21} & n_{22} & n_{23} \end{bmatrix} \times \begin{bmatrix} m_{11} & m_{12} \\ m_{21} & m_{22} \\ m_{31} & m_{32} \end{bmatrix} = \begin{bmatrix} m_{11}n_{11} + m_{21}n_{12} + m_{31}n_{13} & m_{12}n_{11} + m_{22}n_{12} + m_{32}n_{13} \\ m_{11}n_{21} + m_{21}n_{22} + m_{31}n_{23} & m_{12}n_{21} + m_{22}n_{22} + m_{32}n_{23} \end{bmatrix}$$

Review Video: Matrices: The Basics
Visit mometrix.com/academy and enter code: 516658

Matrix Systems

Solving Systems of Equations

Matrices can be used to represent the coefficients of a system of linear equations and can be very useful in solving those systems. Take for instance three equations with three variables where all a, b, c, and d are known constants:

$$a_1x + b_1y + c_1z = d_1$$
$$a_2x + b_2y + c_2z = d_2$$
$$a_3x + b_3y + c_3z = d_3$$

To solve this system, define three matrices:

$$A = \begin{bmatrix} a_1 & b_1 & c_1 \\ a_2 & b_2 & c_2 \\ a_3 & b_3 & c_3 \end{bmatrix}; \; D = \begin{bmatrix} d_1 \\ d_2 \\ d_3 \end{bmatrix}; \; X = \begin{bmatrix} x \\ y \\ z \end{bmatrix}$$

The three equations in our system can be fully represented by a single matrix equation:

$$AX = D$$

We know that the identity matrix times X is equal to X, and we know that any matrix multiplied by its inverse is equal to the identity matrix.

$$A^{-1}AX = IX = X; \text{ thus } X = A^{-1}D$$

Our goal then is to find the inverse of A, or A^{-1}. Once we have that, we can pre-multiply matrix D by A^{-1} (post-multiplying here is an undefined operation) to find matrix X.

Systems of equations can also be solved using the transformation of an augmented matrix in a process similar to that for finding a matrix inverse. Begin by arranging each equation of the system in the following format:

$$\begin{aligned} a_1x + b_1y + c_1z = d_1 \\ a_2x + b_2y + c_2z = d_2 \\ a_3x + b_3y + c_3z = d_3 \end{aligned}$$

Define matrices A and D and combine them into augmented matrix A_a:

$$A = \begin{bmatrix} a_1 & b_1 & c_1 \\ a_2 & b_2 & c_2 \\ a_3 & b_3 & c_3 \end{bmatrix}; \; D = \begin{bmatrix} d_1 \\ d_2 \\ d_3 \end{bmatrix}; A_a = \begin{bmatrix} a_1 & b_1 & c_1 & d_1 \\ a_2 & b_2 & c_2 & d_2 \\ a_3 & b_3 & c_3 & d_3 \end{bmatrix}$$

To solve the augmented matrix and the system of equations, use elementary row operations to form an identity matrix in the first 3×3 section. When this is complete, the values in the last column are the solutions to the system of equations:

$$\begin{bmatrix} 1 & 0 & 0 & x \\ 0 & 1 & 0 & y \\ 0 & 0 & 1 & z \end{bmatrix}$$

If an identity matrix is not possible, the system of equations has no unique solution. Sometimes only a partial solution will be possible. The following are partial solutions you may find:

$\begin{bmatrix} 1 & 0 & k_1 & x_0 \\ 0 & 1 & k_2 & y_0 \\ 0 & 0 & 0 & 0 \end{bmatrix}$ gives the non-unique solution $x = x_0 - k_1z; \; y = y_0 - k_2z$

$\begin{bmatrix} 1 & j_1 & k_1 & x_0 \\ 0 & 0 & 0 & 0 \\ 0 & 0 & 0 & 0 \end{bmatrix}$ gives the non-unique solution $x = x_0 - j_1y - k_1z$

This process can be used to solve systems of equations with any number of variables, but three is the upper limit for practical purposes. Anything more ought to be done with a graphing calculator.

Review Video: Matrices: Data Systems
Visit mometrix.com/academy and enter code: 579763

Precision, Accuracy, and Error

Measurements of physical quantities (e.g., length, area, volume, weight, mass, and time) in the real world are never perfect. Measurements miss the true value, and repeated measurements produce different values. For this reason, fields that depend on good measurement have technical terms, precision and accuracy, that describe how particular ways of measuring a quantity produce good or bad results. Note that these terms apply not to single measurements but to repeated measurements of the same quantity using the same procedure.

Precision describes the consistency of measurements. Measurements that cluster closely together, with little variation from one measurement to the next, have high precision. Measurements that vary a great deal from one measurement to the next have low precision.

Accuracy describes the closeness of measurements to the true value of the quantity being measured or, more technically, the tendency of measurements to cluster around the true value (though we do not usually know the true value; otherwise, we would not bother to measure it).

An analogy from archery may help. An outstanding archer aiming at the center of the target (the true value) produces a tight cluster of arrows around the center (high accuracy, high precision). Under the same circumstances a modestly good archer produces a loose cluster around the center (high accuracy, low precision). If, however, there is a strong crosswind and the archers do nothing to compensate for it, the outstanding archer's shots will cluster tightly around a point away from the center (low accuracy, high precision); and the modestly good archer's shots will cluster loosely around a point away from the center (low accuracy, low precision).

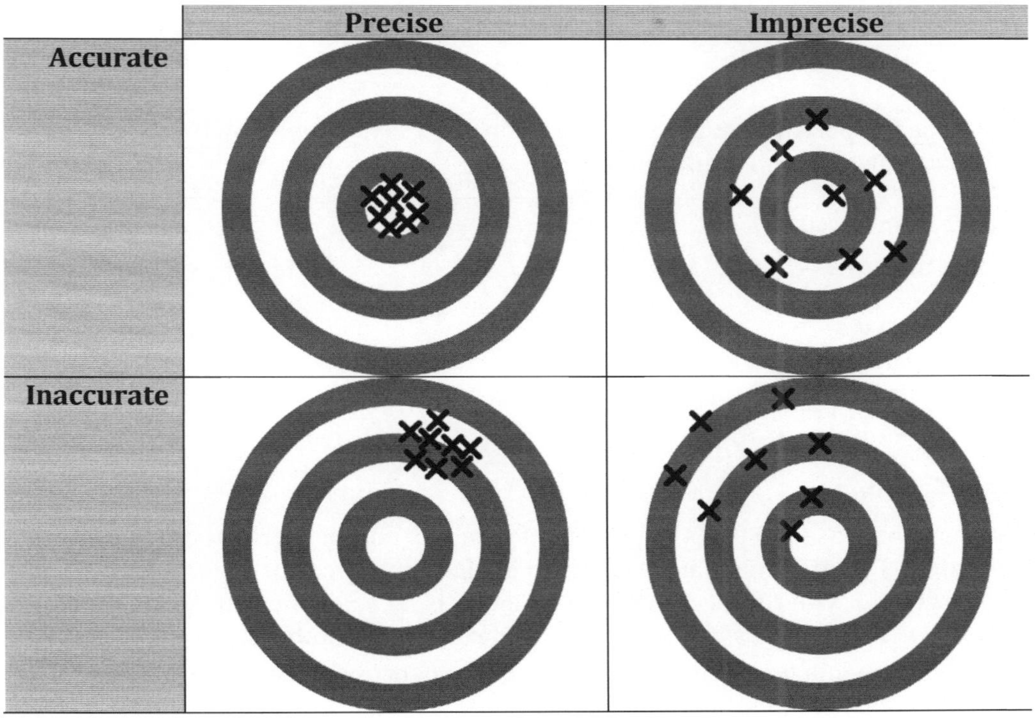

Error in measurement is not a mistake but simply the difference between the true value and the measured value. Since we seldom know the true value, we seldom know the exact error; but sometimes we know a limit on how big the error can be (an error bound), in which case we can use the plus-or-minus sign, $\pm$, to indicate how far the true value might lie from the measured value. For instance, if we measure the length of a metal bar as 25.6 cm and we know the error in our measurement is no more than 0.03 cm (some sources call this **approximate error**), then we might report the length of the bar as 25.6 ± 0.03 cm, which means it lies between 25.57 cm and 25.63 cm.

A particularly important application of this comes from the necessity of rounding whenever we make a measurement. For instance, if we measure the length of a pencil using a ruler marked off in whole inches (no fractions), then we must round our measurement to the nearest inch. Thus, if we report the length as 8 inches, it means that the actual value lies somewhere between 7.5 inches and 8.5 inches, which is to say 8 ± 0.5 inches. This built-in error bound due to rounding is called the **maximum possible error**, and it is always half the magnitude of the smallest unit used in the measurement. Determining the smallest unit can be somewhat subtle, involving the rules for significant figures commonly used in physics, chemistry, and other fields.

Review Video: Precision, Accuracy, and Error
Visit mometrix.com/academy and enter code: 520377

Rectangular and Polar Coordinates

Rectangular coordinates are those that lie on the square grids of the Cartesian plane. They should be quite familiar to you. The polar coordinate system is based on a circular graph, rather than the square grid of the Cartesian system. Points in the polar coordinate system are in the format (r, θ), where r is the distance from the origin (think radius of the circle) and θ is the smallest positive angle (moving counterclockwise around the circle) made with the positive horizontal axis.

To convert a point from rectangular (x, y) format to polar (r, θ) format, use the formula (x, y) to $(r, \theta) \Rightarrow r = \sqrt{x^2 + y^2}$; $\theta = \arctan\frac{y}{x}$ when $x \neq 0$.

If x is positive, use the positive square root value for r. If x is negative, use the negative square root value for r. If $x = 0$, use the following rules:

- If $y = 0$, then $\theta = 0$.
- If $y > 0$, then $\theta = \frac{\pi}{2}$.
- If $y < 0$, then $\theta = \frac{3\pi}{2}$.

To convert a point from polar (r, θ) format to rectangular (x, y) format, use the formula (r, θ) to $(x, y) \Rightarrow x = r\cos\theta$; $y = r\sin\theta$.

De Moivre's Theorem

De Moivre's theorem is used to find the powers of complex numbers (numbers that contain the imaginary number i) written in polar form. Given a trigonometric expression that contains i, such as $z = r\cos x + ir\sin x$, where r is a real number and x is an angle measurement in polar form, use the formula $z^n = r^n(\cos nx + i\sin nx)$, where r and n are real numbers, x is the angle measure in polar form, and i is the imaginary number $i = \sqrt{-1}$. The expression $\cos x + i\sin x$ can be written cis x, making the formula appear in the format $z^n = r^n \text{ cis } nx$.

Note that De Moivre's theorem is only for angles in polar form. If you are given an angle in degrees, you must convert to polar form before using the formula.

Chapter Quiz

Ready to see how well you retained what you just read? Scan the QR code to go directly to the chapter quiz interface for this study guide. If you're using a computer, simply visit the online resources page at **mometrix.com/resources719/cchsalgebra** and click the Chapter Quizzes link.

Algebra

Linear Expressions

Terms and Coefficients

Mathematical expressions consist of a combination of one or more values arranged in terms that are added together. As such, an expression could be just a single number, including zero. A **variable term** is the product of a real number, also called a **coefficient**, and one or more variables, each of which may be raised to an exponent. Expressions may also include numbers without a variable, called **constants** or **constant terms**. The expression $6s^2$, for example, is a single term where the coefficient is the real number 6 and the variable term is s^2. Note that if a term is written as simply a variable to some exponent, like t^2, then the coefficient is 1, because $t^2 = 1t^2$.

Linear Expressions

A **single variable linear expression** is the sum of a single variable term, where the variable has no exponent, and a constant, which may be zero. For instance, the expression $2w + 7$ has $2w$ as the variable term and 7 as the constant term. It is important to realize that terms are separated by addition or subtraction. Since an expression is a sum of terms, expressions such as $5x - 3$ can be written as $5x + (-3)$ to emphasize that the constant term is negative. A real-world example of a single variable linear expression is the perimeter of a square, four times the side length, often expressed: $4s$.

In general, a **linear expression** is the sum of any number of variable terms so long as none of the variables have an exponent and none of the terms have two variables multiplied together. For example, $3m + 8n - \frac{1}{4}p + 5.5q - 1$ is a linear expression, but $3y^3$ and $5xy$ are not. In the same way, the expression for the perimeter of a general triangle $(a + b + c)$ is linear, but the expression for the area of a square (s^2) is not.

Linear Equations

Equations like $5x = 100$ and $8x - 120 = 200$ and $6x + 4y = 240$ are **linear equations.** Linear equations are named based off the number of distinct variables they include. For example, the equation $3x + 30 = 8x$ is a **one-variable linear equation** because it involves only the single variable x. It does not matter that x appears more than once. Any equations that can be written as $ax + b = 0$, where $a \neq 0$, falls into this category. Furthermore, the equation $3x - 5y = 14 + 9y$ is a **two-variable linear equation** because it involves the two variables x and y. The equation $7x + 8y - 12z + 14w = 56$ is a linear equation in four variables.

Satisfying the Equation

When given a one-variable linear equation, the goal is typically to solve it. This means that we want to find the number that makes the equation true if we substitute it for the variable. That number is the **solution,** or root, of the equation. For instance, the equation $5x = 10$ has the solution $x = 2$. This is true because when 2 is substituted for x, the result is $5 \cdot 2 = 10$, which is true. On the other hand, $x = 6$ can not be a solution because $5 \cdot 6 \neq 10$, so it is false. Two equations with the same solution are **equivalent equations**. For example, the equations $5x = 10$ and $5x + 3 = 13$ are equivalent because both have the same solution of $x = 2$.

DETERMINING A SOLUTION SET

The **solution set** is the set of all solutions of an equation. In the previous example, the solution set would be 2. Solutions to a linear equation in two variables consist of pairs of numbers. For instance, the equation $6x + 4y = 240$ has the solution $x = 20$ and $y = 30$ since $6 \cdot 20 + 4 \cdot 30 = 240$ is true. We can write this solution as the ordered pair (20,30) and plot it as a point on the coordinate plane. Such equations usually have infinitely many solutions; and if we plot the points for all these solutions we get a line, which is a picture of all the solutions. We call this **graphing the equation**. When an equation has no true solutions, it is referred to as an **empty set**.

LINEAR EQUATION FORMS

Linear equations can be written many ways. Below is a list of some forms linear equations can take:

- **Standard Form**: $Ax + By = C$; the slope is $\frac{-A}{B}$ and the y-intercept is $\frac{C}{B}$
- **Slope Intercept Form**: $y = mx + b$, where m is the slope and b is the y-intercept
- **Point-Slope Form**: $y - y_1 = m(x - x_1)$, where m is the slope and (x_1, y_1) is a point on the line
- **Two-Point Form**: $\frac{y-y_1}{x-x_1} = \frac{y_2-y_1}{x_2-x_1}$, where (x_1, y_1) and (x_2, y_2) are two points on the given line
- **Intercept Form**: $\frac{x}{x_1} + \frac{y}{y_1} = 1$, where $(x_1, 0)$ is the point at which a line intersects the x-axis, and $(0, y_1)$ is the point at which the same line intersects the y-axis

Review Video: Slope-Intercept and Point-Slope Forms
Visit mometrix.com/academy and enter code: 113216

Review Video: Converting Between Standard and Slope-Intercept Forms
Visit mometrix.com/academy and enter code: 982828

Review Video: Linear Equations Basics
Visit mometrix.com/academy and enter code: 793005

Solving Equations

MANIPULATING EQUATIONS

LIKE TERMS

Like terms are terms in an equation that have the same variable, regardless of whether they also have the same coefficient. This includes terms that *lack* a variable; all constants (i.e., numbers without variables) are considered like terms. If the equation involves terms with a variable raised to different powers, the like terms are those that have the variable raised to the same power.

For example, consider the equation $x^2 + 3x + 2 = 2x^2 + x - 7 + 2x$. In this equation, 2 and –7 are like terms; they are both constants. The terms $3x$, x, and $2x$ are like terms, they all include the variable x raised to the first power. The terms x^2 and $2x^2$ are like terms, they both include the variable x, raised to the second power. The terms $2x$ and $2x^2$ are not like terms; although they both involve the variable x, the variable is not raised to the same power in both terms. The fact that they have the same coefficient, 2, is not relevant.

Review Video: Rules for Manipulating Equations
Visit mometrix.com/academy and enter code: 838871

Carrying Out the Same Operation on Both Sides of an Equation

When solving an equation, the general procedure is to carry out a series of operations on both sides of an equation, choosing operations that simplify the equation when doing so. The reason why the same operation must be carried out on both sides of the equation is because that leaves the meaning of the equation unchanged, and yields a result that is equivalent to the original equation. This would not be the case if we carried out an operation on one side of an equation and not the other. Consider what an equation means: it is a statement that two values or expressions are equal. If we carry out the same operation on both sides of the equation—add 3 to both sides, for example—then the two sides of the equation are changed in the same way, and so remain equal. If we do that to only one side of the equation—add 3 to one side but not the other—then that wouldn't be true; if we change one side of the equation but not the other then the two sides are no longer equal.

Combining Like Terms

Combining like terms refers to adding or subtracting like terms—terms with the same variable—and therefore reducing sets of like terms to a single term. The main advantage of doing this is that it simplifies the equation. Often, combining like terms can be done as the first step in solving an equation, though it can also be done later, such as after distributing terms in a product.

For example, consider the equation $2(x + 3) + 3(2 + x + 3) = -4$. The 2 and the 3 in the second set of parentheses are like terms, and we can combine them, yielding $2(x + 3) + 3(x + 5) = -4$. Now we can carry out the multiplications implied by the parentheses, distributing the outer 2 and 3 accordingly: $2x + 6 + 3x + 15 = -4$. The $2x$ and the $3x$ are like terms, and we can add them together: $5x + 6 + 15 = -4$. Now, the constants 6, 15, and –4 are also like terms, and we can combine them as well: subtracting 6 and 15 from both sides of the equation, we get $5x = -4 - 6 - 15$, or $5x = -25$, which simplifies further to $x = -5$.

Review Video: Solving Equations by Combining Like Terms
Visit mometrix.com/academy and enter code: 668506

Canceling Terms on Opposite Sides of an Equation

Two terms on opposite sides of an equation can be canceled if and only if they *exactly* match each other. They must have the same variable raised to the same power and the same coefficient. For example, in the equation $3x + 2x^2 + 6 = 2x^2 - 6$, $2x^2$ appears on both sides of the equation and can be canceled, leaving $3x + 6 = -6$. The 6 on each side of the equation *cannot* be canceled, because it is added on one side of the equation and subtracted on the other. While they cannot be canceled, however, the 6 and –6 are like terms and can be combined, yielding $3x = -12$, which simplifies further to $x = -4$.

It's also important to note that the terms to be canceled must be independent terms and cannot be part of a larger term. For example, consider the equation $2(x + 6) = 3(x + 4) + 1$. We cannot cancel the x's, because even though they match each other they are part of the larger terms $2(x + 6)$ and $3(x + 4)$. We must first distribute the 2 and 3, yielding $2x + 12 = 3x + 12 + 1$. Now we see that the terms with the x's do not match, but the 12s do, and can be canceled, leaving $2x = 3x + 1$, which simplifies to $x = -1$.

Isolating Variables

To isolate a variable means to manipulate the equation so that the variable appears by itself on one side of the equation, and does not appear at all on the other side. Generally, an equation or inequality is considered to be solved once the variable is isolated and the other side of the equation or inequality is simplified as much as possible. In the case of a two-variable equation or inequality, only one variable needs to be isolated; it will not usually be possible to simultaneously isolate both variables.

For a linear equation—an equation in which the variable only appears raised to the first power—isolating a variable can be done by first moving all the terms with the variable to one side of the equation and all other terms to the other side. (*Moving* a term really means adding the inverse of the term to both sides; when a term is *moved* to the other side of the equation its sign is flipped.) Then combine like terms on each side. Finally, divide both sides by the coefficient of the variable, if applicable. The steps need not necessarily be done in this order, but this order will always work.

Review Video: Solving Equations for Specific Variables
Visit mometrix.com/academy and enter code: 130695

Review Video: Solving Equations Involving Algebraic Fractions
Visit mometrix.com/academy and enter code: 237770

Review Video: Solving One-Step Equations
Visit mometrix.com/academy and enter code: 777004

Solving One-Variable Linear Equations

Equations with One Solution (the Usual Case)

To solve a one-variable linear equation, we use the techniques above to isolate the variable.

1. If any coefficients or constants are fractions, it is often helpful first to multiply both sides of the equation by the least common denominator (of all fractions) to clear the fractions.
2. Simplify both sides of the equation by combining any like terms.
3. Put all terms with the variable on one side of the equation and all constant terms on the other side, by adding or subtracting the same terms on both sides of the equation.
4. Divide both sides by the coefficient of the variable (or multiply both sides by its reciprocal).
5. When we have a value for the variable, we can check it by substituting the value into the original equation to make sure it produces a true result.

Consider the following example for solving the equation $\frac{2}{3}x + 8 = 14$:

$3 \cdot \left(\frac{2}{3}x + 8\right) = 3 \cdot 14$	Clear fractions by multiplying both sides by 3.
$2x + 24 = 42$	Simplify, remembering to apply the distributive property.
$2x + 24 - 24 = 42 - 24$	Subtract 24 from both sides to isolate $2x$.
$2x = 18$	Simplify by combining like terms.
$\frac{2x}{2} = \frac{18}{2}$	Divide both sides by 2 to isolate x.
$x = 9$	Simplify

Finally, we check this answer by substituting $x = 9$ into the original equation to make sure we get a true result.

$$\frac{2}{3}x + 8 = \frac{2}{3}(9) + 8 = 6 + 8 = 14$$

This is correct, so the value of x is 9.

Review Video: Solving Equations Using the Distributive Property
Visit mometrix.com/academy and enter code: 765499

Equations with More Than One Solution

Some types of non-linear equations, such as equations involving squares of variables, may have more than one solution. For example, the equation $x^2 = 4$ has two solutions: 2 and –2. Equations with absolute values can also have multiple solutions: $|x| = 1$ has the solutions $x = 1$ and $x = -1$.

It is possible for a linear equation to have more than one solution but only if the equation is true regardless of the value of the variable. We call such an equation an **identity**. In this case, the equation has infinitely many solutions, because every possible value of the variable is a solution. We discover that a linear equation is an identity when our attempts to isolate the variable cause the variable to disappear, leaving a *true* equation involving only constants. For example, consider the equation $2(3x + 5) = x + 5(x + 2)$. Distributing, we get $6x + 10 = x + 5x + 10$; combining like terms gives $6x + 10 = 6x + 10$, and the $6x$-terms cancel to leave $10 = 10$. This is clearly true, so the original equation is an identity. We could also cancel the 10's leaving $0 = 0$, which is also is clearly true—in general if both sides of the equation can be reduced to match one another exactly, the original equation is an identity.

Equations with No Solution

Some types of non-linear equations, such as equations involving squares of variables, may have no solution. For example, the equation $x^2 = -2$ has no solutions in the real numbers because the square of a real number must be positive. Similarly, $|x| = -1$ has no solution because the absolute value of a number is always positive.

It is also possible for a linear equation to have no solution. We call such an equation a **contradiction**. We discover that a linear equation is a contradiction when our attempts to isolate the variable cause the variable to disappear, leaving a *false* equation involving only constants. For example, the equation $2(x + 3) + x = 3x$ has no solution. We can see this by trying to solve it: first we distribute, leaving $2x + 6 + x = 3x$. Combining like terms gives us $3x + 6 = 3x$, and cancelling the term $3x$ on both sides leaves us with $6 = 0$. This is clearly false, so the original equation is a contradiction, having no solutions.

Features of Equations That Require Special Treatment

A linear equation is an equation in which variables only appear by themselves: not multiplied together, not with exponents other than one, and not inside absolute value signs or any other functions. For example, the equation $x + 1 - 3x = 5 - x$ is a linear equation; while x appears multiple times, it never appears with an exponent other than one, or inside any function. The two-variable equation $2x - 3y = 5 + 2x$ is also a linear equation. In contrast, the equation $x^2 - 5 = 3x$ is *not* a linear equation, because it involves the term x^2. The equation $\sqrt{x} = 5$ is not linear, because it involves a square root. The equation $(x - 1)^2 = 4$ is not linear because even though there's no exponent on the x directly, it appears as part of an expression that is squared. The two-variable equation $x + xy - y = 5$ is not linear because it includes the term xy, where two variables are multiplied together.

As we see above, linear equations can always be solved (or shown to have no solution) by combining like terms and performing simple operations on both sides of the equation. Some non-linear equations can be solved by similar methods, but others may require more advanced methods of solution, if they can be solved analytically at all.

Solving Equations Involving Roots

In an equation involving roots, the first step is to isolate the term with the root, if possible, and then raise both sides of the equation to the appropriate power to eliminate it. Consider an example equation, $2\sqrt{x+1} - 1 = 3$. In this case, begin by adding 1 to both sides, yielding $2\sqrt{x+1} = 4$, and then dividing both sides by 2, yielding $\sqrt{x+1} = 2$. Now square both sides, yielding $x + 1 = 4$. Finally, subtracting 1 from both sides yields $x = 3$.

Squaring both sides of an equation (or raising both sides to any *even* power) may, however, yield a spurious solution—a solution to the squared equation that is *not* a solution of the original equation. It's therefore necessary to plug the solution back into the original equation to make sure it works. In this case, it does: $2\sqrt{3+1} - 1 = 2\sqrt{4} - 1 = 2(2) - 1 = 4 - 1 = 3$.

The same procedure applies for other roots as well. For example, given the equation $3 + \sqrt[3]{2x} = 5$, we can first subtract 3 from both sides, yielding $\sqrt[3]{2x} = 2$ and isolating the root. Raising both sides to the third power yields $2x = 2^3$; i.e., $2x = 8$. We can now divide both sides by 2 to get $x = 4$.

Review Video: Solving Equations Involving Roots
Visit mometrix.com/academy and enter code: 297670

Solving Equations with Exponents

In solving an equation with powers of a variable, sometimes it is possible to eliminate all but one term involving the variable. In that case, we can isolate the power of the variable and then take the appropriate root of both sides to eliminate the exponent. For instance, for the equation $2x^3 + 17 = 5x^3 - 7$, we can subtract $5x^3$ from both sides to get $-3x^3 + 17 = -7$, and then subtract 17 from both sides to get $-3x^3 = -24$. Finally, we can divide both sides by –3 to get $x^3 = 8$. Since this isolates the cube of the variable, we can take the cube root of both sides to get $x = \sqrt[3]{8} = 2$.

One important but often overlooked point is that equations with an exponent greater than 1 may have more than one answer. The solution to $x^2 = 9$ isn't simply $x = 3$; it's $x = \pm 3$ (that is, $x = 3$ or $x = -3$). For a slightly more complicated example, consider the equation $(x-1)^2 - 1 = 3$. Adding 1 to both sides yields $(x-1)^2 = 4$; taking the square root of both sides yields $x - 1 = 2$. We can then add 1 to both sides to get $x = 3$. However, there's a second solution. We also have the possibility that $x - 1 = -2$, in which case $x = -1$. Both $x = 3$ and $x = -1$ are valid solutions, as can be verified by substituting them both into the original equation.

Review Video: Solving Equations with Exponents
Visit mometrix.com/academy and enter code: 514557

Review Video: Adding and Subtracting with Exponents
Visit mometrix.com/academy and enter code: 875756

Solving Equations with Absolute Values

When solving an equation with an absolute value, the first step is to isolate the absolute value term. We then consider two possibilities: when the expression inside the absolute value is positive or when it is negative. In the former case, the expression in the absolute value equals the expression on the other side of the equation; in the latter, it equals the additive inverse of that expression—the expression times negative one. We consider each case separately and finally check for spurious solutions.

For instance, consider solving $|2x - 1| + x = 5$ for x. We can first isolate the absolute value by moving the x to the other side: $|2x - 1| = -x + 5$. Now, we have two possibilities. First, that $2x - 1$ is positive, and hence $2x - 1 = -x + 5$. Rearranging and combining like terms yields $3x = 6$, and hence $x = 2$. The other possibility is that $2x - 1$ is negative, and hence $2x - 1 = -(-x + 5) = x - 5$. In this case, rearranging and combining like terms yields $x = -4$. Substituting $x = 2$ and $x = -4$ back into the original equation, we see that they are both valid solutions.

Note that the absolute value of a sum or difference applies to the sum or difference as a whole, not to the individual terms; in general, $|2x - 1|$ is not equal to $|2x + 1|$ or to $|2x| - 1$.

Review Video: Solving Absolute Value Equations
Visit mometrix.com/academy and enter code: 501208

Extraneous Solutions

An **extraneous solution** may arise when we square both sides of an equation (or raise both sides to an even power) as a step in solving it or under certain other operations on the equation. It is a solution to the squared or otherwise modified equation that is *not* a solution of the original equation. To identify an extraneous solution, it's useful when you solve an equation involving roots or absolute values to plug the solution back into the original equation to make sure it's valid.

Two-Variable Equations

Similar to methods for a one-variable equation, solving a two-variable equation involves isolating a variable: manipulating the equation so that a variable appears by itself on one side of the equation, and not at all on the other side. However, in a two-variable equation, you will usually only be able to isolate one of the variables; the other variable may appear on the other side along with constant terms, or with exponents or other functions. If an equation has multiple variables, the problem should tell you which variable to isolate.

Review Video: Solving Equations with Variables on Both Sides
Visit mometrix.com/academy and enter code: 402497

Graphing Equations

Graphical Solutions to Equations

When equations are shown graphically, they are usually shown on a **Cartesian coordinate plane**. The Cartesian coordinate plane consists of two number lines placed perpendicular to each other and intersecting at the zero point, also known as the origin. The horizontal number line is known as the x-axis, with positive values to the right of the origin, and negative values to the left of the origin. The vertical number line is known as the y-axis, with positive values above the origin, and negative values below the origin. Any point on the plane can be identified by an ordered pair in the form (x, y), called coordinates. The x-value of the coordinate is called the abscissa, and the y-value of the coordinate is called the ordinate. The two number lines divide the plane into **four quadrants**: I, II, III, and IV.

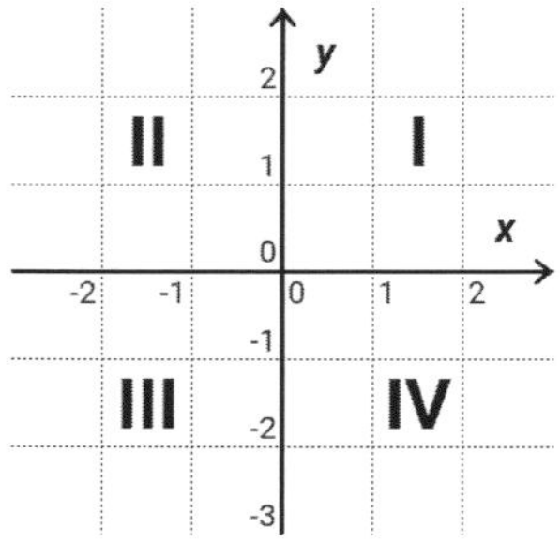

Note that in quadrant I $x > 0$ and $y > 0$, in quadrant II $x < 0$ and $y > 0$, in quadrant III $x < 0$ and $y < 0$, and in quadrant IV $x > 0$ and $y < 0$.

Recall that if the value of the slope of a line is positive, the line slopes upward from left to right. If the value of the slope is negative, the line slopes downward from left to right. If the y-coordinates are the same for two points on a line, the slope is 0 and the line is a **horizontal line**. If the x-coordinates are the same for two points on a line, there is no slope and the line is a **vertical line**. Two or more lines that have equivalent slopes are **parallel lines**. **Perpendicular lines** have slopes that are negative reciprocals of each other, such as $\frac{a}{b}$ and $\frac{-b}{a}$.

Review Video: Cartesian Coordinate Plane and Graphing
Visit mometrix.com/academy and enter code: 115173

Graphing Equations in Two Variables

One way of graphing an equation in two variables is to plot enough points to get an idea for its shape and then draw the appropriate curve through those points. A point can be plotted by substituting in a value for one variable and solving for the other. If the equation is linear, we only need two points and can then draw a straight line between them.

For example, consider the equation $y = 2x - 1$. This is a linear equation—both variables only appear raised to the first power—so we only need two points. When $x = 0$, $y = 2(0) - 1 = -1$. When $x = 2$, $y = 2(2) - 1 = 3$. We can therefore choose the points $(0, -1)$ and $(2, 3)$, and draw a line between them:

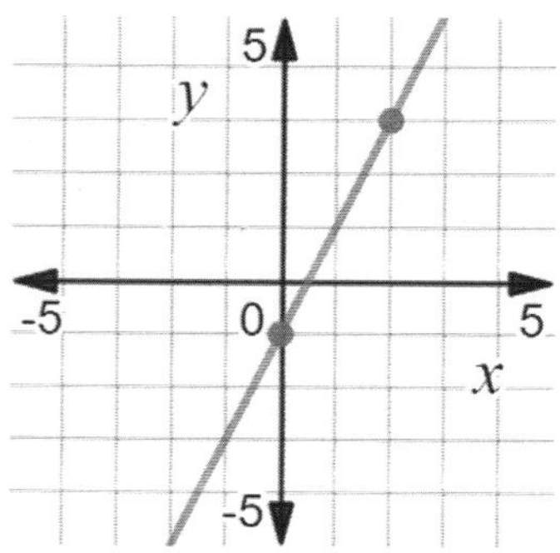

Inequalities

Commonly in algebra and other upper-level fields of math you find yourself working with mathematical expressions that do not equal each other. The statement comparing such expressions with symbols such as $<$ (less than) or $>$ (greater than) is called an *inequality*. An example of an inequality is $7x > 5$. To solve for x, simply divide both sides by 7 and the solution is shown to be $x > \frac{5}{7}$. Graphs of the solution set of inequalities are represented on a number line. Open circles are used to show that an expression approaches a number but is never quite equal to that number.

Review Video: Solving One-Step Inequalities
Visit mometrix.com/academy and enter code: 229684

Review Video: Solving Multi-Step Inequalities
Visit mometrix.com/academy and enter code: 347842

Review Video: Solving Inequalities Using All 4 Basic Operations
Visit mometrix.com/academy and enter code: 401111

Types of Inequalities

Conditional inequalities are those with certain values for the variable that will make the condition true and other values for the variable where the condition will be false. **Absolute inequalities** can have any real number as the value for the variable to make the condition true, while there is no real number value for the variable that will make the condition false. Solving inequalities is done by following the same rules for solving equations with the exception that when multiplying or dividing by a negative number the direction of the inequality sign must be flipped or reversed. **Double inequalities** are situations where two inequality statements apply to the same variable expression. Example: $-c < ax + b < c$.

Review Video: Conditional and Absolute Inequalities
Visit mometrix.com/academy and enter code: 980164

Solving Inequalities

Determining Solutions to Inequalities

To determine whether a coordinate is a solution of an inequality, you can substitute the values of the coordinate into the inequality, simplify, and check whether the resulting statement holds true. For instance, to determine whether $(-2,4)$ is a solution of the inequality $y \geq -2x + 3$, substitute the values into the inequality, $4 \geq -2(-2) + 3$. Simplify the right side of the inequality and the result is $4 \geq 7$, which is a false statement. Therefore, the coordinate is not a solution of the inequality. You can also use this method to determine which part of the graph of an inequality is shaded. The graph of $y \geq -2x + 3$ includes the solid line $y = -2x + 3$ and, since it excludes the point $(-2,4)$ to the left of the line, it is shaded to the right of the line.

Review Video: Graphing Linear Inequalities
Visit mometrix.com/academy and enter code: 439421

Review Video: Graphing Solutions to Inequalities
Visit mometrix.com/academy and enter code: 391281

Flipping Inequality Signs

When given an inequality, we can always turn the entire inequality around, swapping the two sides of the inequality and changing the inequality sign. For instance, $x + 2 > 2x - 3$ is equivalent to $2x - 3 < x + 2$. Aside from that, normally the inequality does not change if we carry out the same operation on both sides of the inequality. There is, however, one principal exception: if we *multiply* or *divide* both sides of the inequality by a *negative number*, the inequality is flipped. For example, if we take the inequality $-2x < 6$ and divide both sides by –2, the inequality flips and we are left with $x > -3$. This *only* applies to multiplication and division, and only with negative numbers. Multiplying or dividing both sides by a positive number, or adding or subtracting any number regardless of sign, does not flip the inequality. Another special case that flips the inequality sign is when reciprocals are used. For instance, $3 > 2$ but the relation of the reciprocals is $\frac{1}{3} < \frac{1}{2}$.

Compound Inequalities

A **compound inequality** is an equality that consists of two inequalities combined with *and* or *or*. The two components of a proper compound inequality must be of opposite type: that is, one must be greater than (or greater than or equal to), the other less than (or less than or equal to). For instance, "$x + 1 < 2$ or $x + 1 > 3$" is a compound inequality, as is "$2x \geq 4$ and $2x \leq 6$." An *and* inequality can be written more compactly by having one inequality on each side of the common part: "$2x \geq 1$ and $2x \leq 6$," can also be written as $1 \leq 2x \leq 6$.

In order for the compound inequality to be meaningful, the two parts of an *and* inequality must overlap; otherwise, no numbers satisfy the inequality. On the other hand, if the two parts of an *or* inequality overlap, then *all* numbers satisfy the inequality and as such the inequality is usually not meaningful.

Solving a compound inequality requires solving each part separately. For example, given the compound inequality "$x + 1 < 2$ or $x + 1 > 3$," the first inequality, $x + 1 < 2$, reduces to $x < 1$, and the second part, $x + 1 > 3$, reduces to $x > 2$, so the whole compound inequality can be written as "$x < 1$ or $x > 2$." Similarly, $1 \leq 2x \leq 6$ can be solved by dividing each term by 2, yielding $\frac{1}{2} \leq x \leq 3$.

Review Video: Compound Inequalities
Visit mometrix.com/academy and enter code: 786318

Solving Inequalities Involving Absolute Values

To solve an inequality involving an absolute value, first isolate the term with the absolute value. Then proceed to treat the two cases separately as with an absolute value equation, but flipping the inequality in the case where the expression in the absolute value is negative (since that essentially involves multiplying both sides by –1.) The two cases are then combined into a compound inequality; if the absolute value is on the greater side of the inequality, then it is an *or* compound inequality, if on the lesser side, then it's an *and*.

Consider the inequality $2 + |x - 1| \geq 3$. We can isolate the absolute value term by subtracting 2 from both sides: $|x - 1| \geq 1$. Now, we're left with the two cases $x - 1 \geq 1$ or $x - 1 \leq -1$: note that in the latter, negative case, the inequality is flipped. $x - 1 \geq 1$ reduces to $x \geq 2$, and $x - 1 \leq -1$ reduces to $x \leq 0$. Since in the inequality $|x - 1| \geq 1$ the absolute value is on the greater side, the two cases combine into an *or* compound inequality, so the final, solved inequality is "$x \leq 0$ or $x \geq 2$."

Review Video: Solving Absolute Value Inequalities
Visit mometrix.com/academy and enter code: 997008

Solving Inequalities Involving Square Roots

Solving an inequality with a square root involves two parts. First, we solve the inequality as if it were an equation, isolating the square root and then squaring both sides of the equation. Second, we restrict the solution to the set of values of x for which the value inside the square root sign is non-negative.

For example, in the inequality, $\sqrt{x-2}+1<5$, we can isolate the square root by subtracting 1 from both sides, yielding $\sqrt{x-2}<4$. Squaring both sides of the inequality yields $x-2<16$, so $x<18$. Since we can't take the square root of a negative number, we also require the part inside the square root to be non-negative. In this case, that means $x-2\geq 0$. Adding 2 to both sides of the inequality yields $x\geq 2$. Our final answer is a compound inequality combining the two simple inequalities: $x\geq 2$ and $x<18$, or $2\leq x<18$.

Note that we only get a compound inequality if the two simple inequalities are in opposite directions; otherwise, we take the one that is more restrictive.

The same technique can be used for other even roots, such as fourth roots. It is *not*, however, used for cube roots or other odd roots—negative numbers *do* have cube roots, so the condition that the quantity inside the root sign cannot be negative does not apply.

Review Video: Solving Inequalities Involving Square Roots
Visit mometrix.com/academy and enter code: 800288

Special Circumstances

Sometimes an inequality involving an absolute value or an even exponent is true for all values of x, and we don't need to do any further work to solve it. This is true if the inequality, once the absolute value or exponent term is isolated, says that term is greater than a negative number (or greater than or equal to zero). Since an absolute value or a number raised to an even exponent is *always* non-negative, this inequality is always true.

Graphing Inequalities

Graphing Simple Inequalities

To graph a simple inequality, we first mark on the number line the value that signifies the end point of the inequality. If the inequality is strict (involves a less than or greater than), we use a hollow circle; if it is not strict (less than or equal to or greater than or equal to), we use a solid circle. We then fill in the part of the number line that satisfies the inequality: to the left of the marked point for less than (or less than or equal to), to the right for greater than (or greater than or equal to).

For example, we would graph the inequality $x < 5$ by putting a hollow circle at 5 and filling in the part of the line to the left:

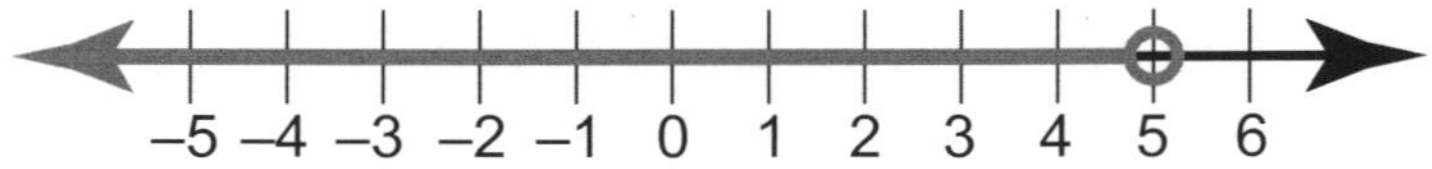

GRAPHING COMPOUND INEQUALITIES

To graph a compound inequality, we fill in both parts of the inequality for an *or* inequality, or the overlap between them for an *and* inequality. More specifically, we start by plotting the endpoints of each inequality on the number line. For an *or* inequality, we then fill in the appropriate side of the line for each inequality. Typically, the two component inequalities do not overlap, which means the shaded part is *outside* the two points. For an *and* inequality, we instead fill in the part of the line that meets both inequalities.

For the inequality "$x \leq -3$ or $x > 4$," we first put a solid circle at –3 and a hollow circle at 4. We then fill the parts of the line *outside* these circles:

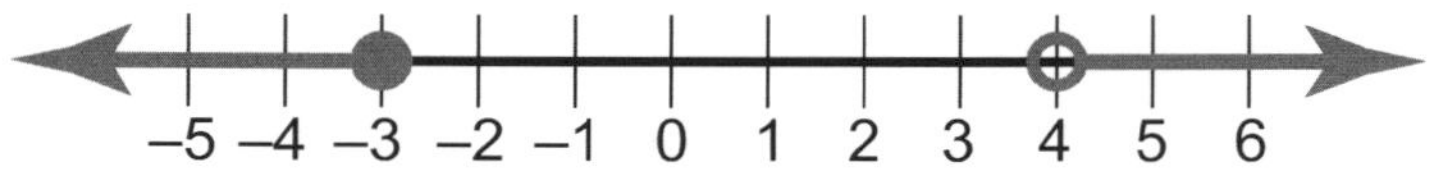

GRAPHING INEQUALITIES INCLUDING ABSOLUTE VALUES

An inequality with an absolute value can be converted to a compound inequality. To graph the inequality, first convert it to a compound inequality, and then graph that normally. If the absolute value is on the greater side of the inequality, we end up with an *or* inequality; we plot the endpoints of the inequality on the number line and fill in the part of the line *outside* those points. If the absolute value is on the smaller side of the inequality, we end up with an *and* inequality; we plot the endpoints of the inequality on the number line and fill in the part of the line *between* those points.

For example, the inequality $|x + 1| \geq 4$ can be rewritten as $x \geq 3$ or $x \leq -5$. We place solid circles at the points 3 and –5 and fill in the part of the line *outside* them:

GRAPHING INEQUALITIES IN TWO VARIABLES

To graph an inequality in two variables, we first graph the border of the inequality. This means graphing the equation that we get if we replace the inequality sign with an equals sign. If the inequality is strict ($>$ or $<$), we graph the border with a dashed or dotted line; if it is not strict ($\geq$ or $\leq$), we use a solid line. We can then test any point not on the border to see if it satisfies the inequality. If it does, we shade in that side of the border; if not, we shade in the other side. As an example, consider $y > 2x + 2$. To graph this inequality, we first graph the border, $y = 2x + 2$. Since it is a strict inequality, we use a dashed line. Then, we choose a test point. This can be any point not on the border; in this case, we will choose the origin, (0,0). (This makes the calculation easy and is generally a good choice unless the border passes through the origin.) Putting this into the original inequality, we get $0 > 2(0) + 2$, i.e., $0 > 2$. This is *not* true, so we shade in the side of the border that does *not* include the point (0,0):

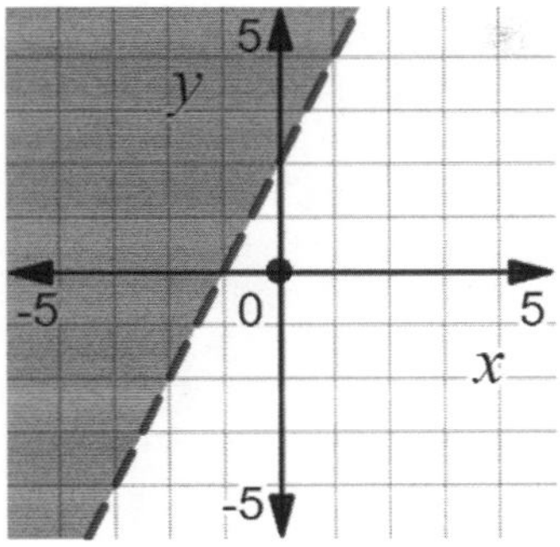

GRAPHING COMPOUND INEQUALITIES IN TWO VARIABLES

One way to graph a compound inequality in two variables is to first graph each of the component inequalities. For an *and* inequality, we then shade in only the parts where the two graphs overlap; for an *or* inequality, we shade in any region that pertains to either of the individual inequalities.

Consider the graph of "$y \geq x - 1$ and $y \leq -x$":

We first shade in the individual inequalities:

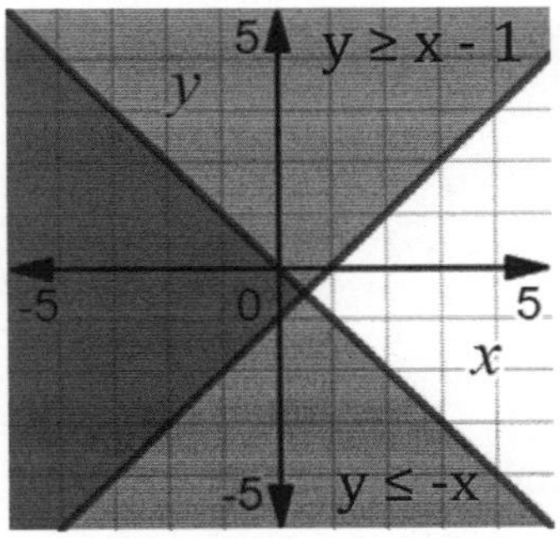

Now, since the compound inequality has an *and*, we only leave shaded the overlap—the part that pertains to *both* inequalities:

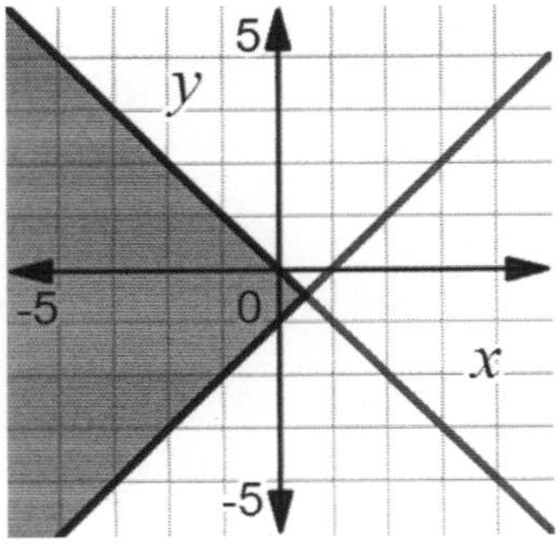

If instead the inequality had been "$y \geq x - 1$ or $y \leq -x$," our final graph would involve the *total* shaded area:

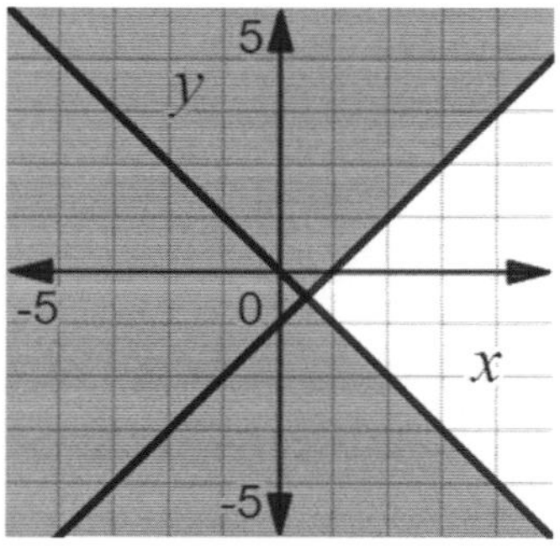

Review Video: Graphing Solutions to Inequalities
Visit mometrix.com/academy and enter code: 391281

Systems of Equations

Solving Systems of Equations

A **system of equations** is a set of simultaneous equations that all use the same variables. A solution to a system of equations must be true for each equation in the system. **Consistent systems** are those with at least one solution. **Inconsistent systems** are systems of equations that have no solution.

Review Video: Solving Systems of Linear Equations
Visit mometrix.com/academy and enter code: 746745

Substitution

To solve a system of linear equations by **substitution**, start with the easier equation and solve for one of the variables. Express this variable in terms of the other variable. Substitute this expression in the other equation and solve for the other variable. The solution should be expressed in the form (x, y). Substitute the values into both of the original equations to check your answer. Consider the following system of equations:

$$x + 6y = 15$$
$$3x - 12y = 18$$

Solving the first equation for x: $x = 15 - 6y$

Substitute this value in place of x in the second equation, and solve for y:

$$3(15 - 6y) - 12y = 18$$
$$45 - 18y - 12y = 18$$
$$30y = 27$$
$$y = \frac{27}{30} = \frac{9}{10} = 0.9$$

Plug this value for y back into the first equation to solve for x:

$$x = 15 - 6(0.9) = 15 - 5.4 = 9.6$$

Check both equations if you have time:

$$\begin{aligned} 9.6 + 6(0.9) &= 15 \\ 9.6 + 5.4 &= 15 \\ 15 &= 15 \end{aligned} \qquad \begin{aligned} 3(9.6) - 12(0.9) &= 18 \\ 28.8 - 10.8 &= 18 \\ 18 &= 18 \end{aligned}$$

Therefore, the solution is $(9.6, 0.9)$.

Review Video: The Substitution Method
Visit mometrix.com/academy and enter code: 565151

Review Video: Substitution and Elimination
Visit mometrix.com/academy and enter code: 958611

ELIMINATION

To solve a system of equations using **elimination**, begin by rewriting both equations in standard form $Ax + By = C$. Check to see if the coefficients of one pair of like variables add to zero. If not, multiply one or both of the equations by a non-zero number to make one set of like variables add to zero. Add the two equations to solve for one of the variables. Substitute this value into one of the original equations to solve for the other variable. Check your work by substituting into the other equation. Now, let's look at solving the following system using the elimination method:

$$\begin{aligned} 5x + 6y &= 4 \\ x + 2y &= 4 \end{aligned}$$

If we multiply the second equation by -3, we can eliminate the y-terms:

$$\begin{aligned} 5x + 6y &= 4 \\ -3x - 6y &= -12 \end{aligned}$$

Add the equations together and solve for x:

$$\begin{aligned} 2x &= -8 \\ x &= \frac{-8}{2} = -4 \end{aligned}$$

Plug the value for x back in to either of the original equations and solve for y:

$$\begin{aligned} -4 + 2y &= 4 \\ y &= \frac{4 + 4}{2} = 4 \end{aligned}$$

Check both equations if you have time:

$$\begin{aligned} 5(-4) + 6(4) &= 4 \\ -20 + 24 &= 4 \\ 4 &= 4 \end{aligned} \qquad \begin{aligned} -4 + 2(4) &= 4 \\ -4 + 8 &= 4 \\ 4 &= 4 \end{aligned}$$

Therefore, the solution is $(-4,4)$.

Review Video: The Elimination Method
Visit mometrix.com/academy and enter code: 449121

GRAPHICALLY

To solve a system of linear equations **graphically**, plot both equations on the same graph. The solution of the equations is the point where both lines cross. If the lines do not cross (are parallel), then there is **no solution**.

For example, consider the following system of equations:

$$y = 2x + 7$$
$$y = -x + 1$$

Since these equations are given in slope-intercept form, they are easy to graph; the y-intercepts of the lines are (0,7) and (0,1). The respective slopes are 2 and –1, thus the graphs look like this:

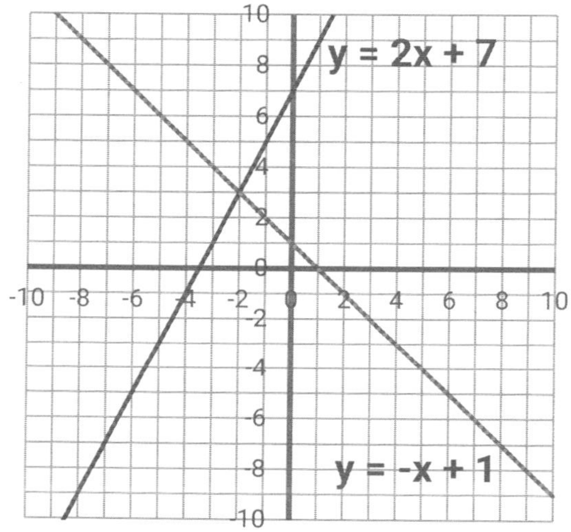

The two lines intersect at the point $(-2,3)$, thus this is the solution to the system of equations.

Solving a system graphically is generally only practical if both coordinates of the solution are integers; otherwise the intersection will lie between gridlines on the graph and the coordinates will be difficult or impossible to determine exactly. It also helps if, as in this example, the equations are in slope-intercept form or some other form that makes them easy to graph. Otherwise, another method of solution (by substitution or elimination) is likely to be more useful.

Review Video: Solving Systems by Graphing
Visit mometrix.com/academy and enter code: 634812

SOLVING SYSTEMS OF EQUATIONS USING THE TRACE FEATURE

Using the trace feature on a calculator requires that you rewrite each equation, isolating the y-variable on one side of the equal sign. Enter both equations in the graphing calculator and plot the graphs simultaneously. Use the trace cursor to find where the two lines cross. Use the zoom feature if necessary to obtain more accurate results. Always check your answer by substituting into the original equations. The trace method is likely to be less accurate than other methods due to the resolution of graphing calculators but is a useful tool to provide an approximate answer.

Advanced Systems of Equations

Solving a System of Equations with a Linear Equation and a Quadratic Equation

Algebraically

Generally, the simplest way to solve a system of equations consisting of a linear equation and a quadratic equation algebraically is through the method of substitution. One possible strategy is to solve the linear equation for y and then substitute that expression into the quadratic equation. After expansion and combining like terms, this will result in a new quadratic equation for x, which, like all quadratic equations, may have zero, one, or two solutions. Plugging each solution for x back into one of the original equations will then produce the corresponding value of y.

For example, consider the following system of equations:

$$x + y = 1$$
$$y = (x + 3)^2 - 2$$

We can solve the linear equation for y to yield $y = -x + 1$. Substituting this expression into the quadratic equation produces $-x + 1 = (x + 3)^2 - 2$. We can simplify this equation:

$$\begin{aligned} -x + 1 &= (x + 3)^2 - 2 \\ -x + 1 &= x^2 + 6x + 9 - 2 \\ -x + 1 &= x^2 + 6x + 7 \\ 0 &= x^2 + 7x + 6 \end{aligned}$$

This quadratic equation can be factored as $(x + 1)(x + 6) = 0$. It therefore has two solutions: $x_1 = -1$ and $x_2 = -6$. Plugging each of these back into the original linear equation yields $y_1 = -x_1 + 1 = -(-1) + 1 = 2$ and $y_2 = -x_2 + 1 = -(-6) + 1 = 7$. Thus, this system of equations has two solutions, $(-1,2)$ and $(-6,7)$.

It may help to check your work by putting each x- and y-value back into the original equations and verifying that they do provide a solution.

Graphically

To solve a system of equations consisting of a linear equation and a quadratic equation graphically, plot both equations on the same graph. The linear equation will, of course, produce a straight line, while the quadratic equation will produce a parabola. These two graphs will intersect at zero, one, or two points; each point of intersection is a solution of the system.

For example, consider the following system of equations:

$$y = -2x + 2$$
$$y = -2x^2 + 4x + 2$$

The linear equation describes a line with a y-intercept of $(0,2)$ and a slope of -2.

To graph the quadratic equation, we can first find the vertex of the parabola: the x-coordinate of the vertex is $h = -\frac{b}{2a} = -\frac{4}{2(-2)} = 1$, and the y-coordinate is $k = -2(1)^2 + 4(1) + 2 = 4$. Thus, the vertex lies at $(1,4)$. To get a feel for the rest of the parabola, we can plug in a few more values of x to find more points; by putting in $x = 2$ and $x = 3$ in the quadratic equation, we find that the points $(2,2)$ and $(3,-4)$ lie on the parabola; by symmetry, so must $(0,2)$ and $(-1,-4)$. We can now plot both equations:

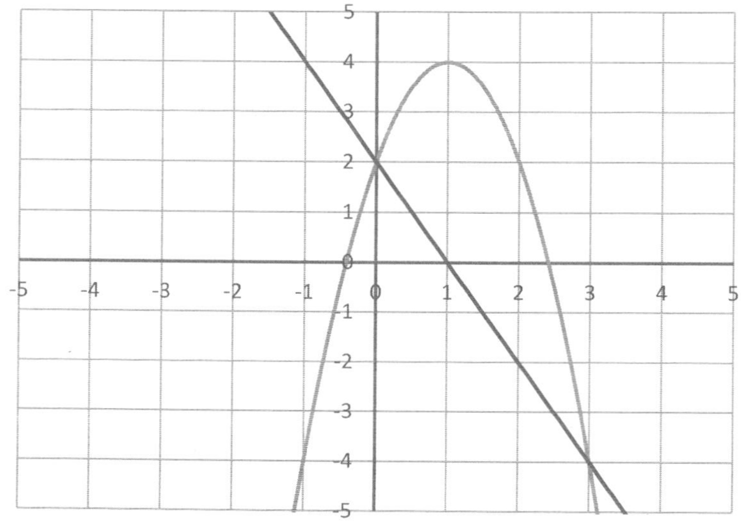

These two curves intersect at the points $(0,2)$ and $(3,-4)$, thus these are the solutions of the equation.

Review Video: **Parabolas**
Visit mometrix.com/academy and enter code: 129187

Review Video: **Vertex of a Parabola**
Visit mometrix.com/academy and enter code: 272300

Review Video: Solving a System of Linear and Quadratic Equations
Visit mometrix.com/academy and enter code: 194870

Midpoint and Distance Formulas

If you know the coordinates of the endpoints of a line segment, you can calculate the midpoint and length of the line segment. Conveniently, the length of the line segment is also the distance between the two endpoints.

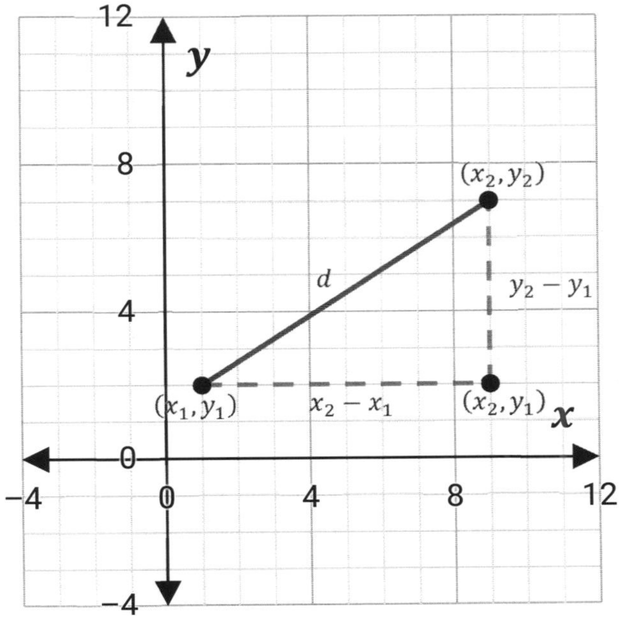

To find the **midpoint** of the line segment with endpoints (x_1, y_1) and (x_2, y_2), average the x-coordinates to get the x-coordinate of the midpoint, and average the y-coordinates to get the y-coordinate of the midpoint. Thus, the **midpoint formula** is:

$$\left(\frac{x_1 + x_2}{2}, \frac{y_1 + y_2}{2}\right)$$

The **distance** between points (x_1, y_1) and (x_2, y_2) is the same as the length of the hypotenuse of a right triangle with the two given points as endpoints, and the two sides of the right triangle parallel to the x-axis and y-axis, respectively. The length of the segment parallel to the x-axis is the difference between the x-coordinates of the two points. The length of the segment parallel to the y-axis is the difference between the y-coordinates of the two points. Use the Pythagorean theorem $a^2 + b^2 = c^2$ or $c = \sqrt{a^2 + b^2}$ to find the distance. Thus, the **distance formula** is:

$$d = \sqrt{(x_2 - x_1)^2 + (y_2 - y_1)^2}$$

Review Video: Calculations Using Points on a Graph
Visit mometrix.com/academy and enter code: 883228

Polynomials

Monomials and Polynomials

A **monomial** is a single constant, variable, or product of constants and variables, such as 7, x, $2x$, or x^3y. There will never be addition or subtraction symbols in a monomial. Like monomials have like variables, but they may have different coefficients. A **polynomial** is a monomial or the result of combining two or more monomials by sums or differences. In a polynomial we call each monomial a **term**. Two terms make a **binomial** (e.g., $2x + 3y$), three terms make a **trinomial** (e.g., $5x^2 - 4x + 9$). The **degree of a monomial** is the sum of the exponents of the variables. The **degree of a polynomial** is the highest degree of any individual term.

Review Video: Polynomials
Visit mometrix.com/academy and enter code: 305005

Simplifying Polynomials

Simplifying polynomials requires combining like terms. The like terms in a polynomial expression are those that have the same variables raised to the same powers. It is often helpful to connect the like terms with arrows or lines in order to separate them from the other monomials. Once you have determined the like terms, you can rearrange the polynomial by placing them together. Remember to include the sign that is in front of each term. Once the like terms are placed together, you can apply each operation and simplify. When adding and subtracting polynomials, only add and subtract the **coefficients**, or the number part; the variable and exponent stay the same.

Adding Polynomials

To add polynomials, you need to add like terms. These terms have the same variable part. For example, the terms $4x^2$ and $3x^2$ both include x^2 terms. To find the sum of like terms, find the sum of the coefficients. Then, keep the same variable part. You can use the distributive property to distribute the plus sign to each term of the polynomial. For example:

$$(4x^2 - 5x + 7) + (3x^2 + 2x + 1) =$$
$$(4x^2 - 5x + 7) + 3x^2 + 2x + 1 =$$
$$(4x^2 + 3x^2) + (-5x + 2x) + (7 + 1) =$$
$$7x^2 - 3x + 8$$

Subtracting Polynomials

To subtract polynomials, you need to subtract like terms. To find the difference of like terms, find the difference of the coefficients. Then, keep the same variable part. You can use the distributive property to distribute the minus sign to each term of the polynomial. For example:

$$(-2x^2 - x + 5) - (3x^2 - 4x + 1) =$$
$$(-2x^2 - x + 5) - 3x^2 + 4x - 1 =$$
$$(-2x^2 - 3x^2) + (-x + 4x) + (5 - 1) =$$
$$-5x^2 + 3x + 4$$

Review Video: Adding and Subtracting Polynomials
Visit mometrix.com/academy and enter code: 124088

Multiplying Polynomials

In general, multiplying polynomials is done by multiplying each term in one polynomial by each term in the other and adding the results. In the specific case for multiplying binomials, there is a useful acronym, FOIL, that can help you make sure to cover each combination of terms. The **FOIL method** for $(Ax + By)(Cx + Dy)$ would be:

F	Multiply the *first* terms of each binomial	$(\overbrace{Ax}^{first} + By)(\overbrace{Cx}^{first} + Dy)$	ACx^2
O	Multiply the *outer* terms	$(\overbrace{Ax}^{outer} + By)(Cx + \overbrace{Dy}^{outer})$	$ADxy$
I	Multiply the *inner* terms	$(Ax + \overbrace{By}^{inner})(\overbrace{Cx}^{inner} + Dy)$	$BCxy$
L	Multiply the *last* terms of each binomial	$(Ax + \overbrace{By}^{last})(Cx + \overbrace{Dy}^{last})$	BDy^2

Then, add up the result of each and combine like terms: $ACx^2 + (AD + BC)xy + BDy^2$.

For example, using the FOIL method on binomials $(x + 2)$ and $(x - 3)$:

$$\begin{aligned}
\text{First:}\quad & (\boxed{x} + 2)(\boxed{x} + (-3)) \rightarrow (x)(x) = x^2 \\
\text{Outer:}\quad & (\boxed{x} + 2)(x + \boxed{(-3)}) \rightarrow (x)(-3) = -3x \\
\text{Inner:}\quad & (x + \boxed{2})(\boxed{x} + (-3)) \rightarrow (2)(x) = 2x \\
\text{Last:}\quad & (x + \boxed{2})(x + \boxed{(-3)}) \rightarrow (2)(-3) = -6
\end{aligned}$$

This results in: $(x^2) + (-3x) + (2x) + (-6)$

Combine like terms: $x^2 + (-3 + 2)x + (-6) = x^2 - x - 6$

Review Video: Multiplying Polynomials
Visit mometrix.com/academy and enter code: 598293

Review Video: Multiplying Terms Using the FOIL Method
Visit mometrix.com/academy and enter code: 854792

Dividing Polynomials

Use long division to divide a polynomial by either a monomial or another polynomial of equal or lesser degree.

When **dividing by a monomial**, divide each term of the polynomial by the monomial.

Review Video: Dividing Monomials
Visit mometrix.com/academy and enter code: 584409

When **dividing by a polynomial**, begin by arranging the terms of each polynomial in order of one variable. You may arrange in ascending or descending order, but be consistent with both polynomials. To get the first term of the quotient, divide the first term of the dividend by the first term of the divisor. Multiply the first term of the quotient by the entire divisor and subtract that product from the dividend. Repeat for the second and successive terms until you either get a remainder of zero or a remainder whose degree is less than the degree of the divisor. If the quotient has a remainder, write the answer as a mixed expression in the form:

$$\text{quotient} + \frac{\text{remainder}}{\text{divisor}}$$

For example, we can evaluate the following expression in the same way as long division:

$$\frac{x^3 - 3x^2 - 2x + 5}{x - 5}$$

$$\begin{array}{r}
x^2 \quad + 2x \quad + 8 \\
x - 5 \overline{\smash{)}\; x^3 - 3x^2 \quad - 2x \quad + 5} \\
-\underline{(x^3 - 5x^2)} \qquad\qquad\quad \\
2x^2 \quad - 2x \qquad\quad \\
-\underline{(2x^2 - 10x)} \qquad\quad \\
8x \quad + 5 \\
-\underline{(8x - 40)} \\
45
\end{array}$$

$$\frac{x^3 - 3x^2 - 2x + 5}{x - 5} = x^2 + 2x + 8 + \frac{45}{x - 5}$$

Review Video: Dividing Polynomials by Monomials
Visit mometrix.com/academy and enter code: 253551

Review Video: Dividing Trinomials by Binomials
Visit mometrix.com/academy and enter code: 651465

When **factoring** a polynomial, first see whether you can factor out a nontrivial greatest common factor (GCF). For example, the trinomial $3x^5 - 18x^4 + 15x^3$ has a GCF of $3x^3$ since the GCF of 3, 18, and 15 is 3 and the GCF of x^5, x^4, and x^3 is x^3. Factoring out the GCF simplifies the expression to $3x^3(x^2 - 6x + 5)$.

To factor a quadratic trinomial (this comes up frequently), first check whether it is a perfect square trinomial (see bulleted list below). If not, see if you can factor it by trial and error by making clever choices of values for a and b (or a, b, c, and d) in the formulas below (this amounts to trying to use the FOIL mnemonic backwards):

$$x^2 + (a + b)x + ab = (x + a)(x + b)$$
$$(ac)x^2 + (ad + bc)x + bd = (ax + b)(cx + d)$$

For instance, you would try to factor the trinomial $x^2 - 6x + 5$ using the equation $x^2 + (a + b)x + ab = (x + a)(x + b)$. This means that you need to find integers a and b such that $a + b = -6$ and $ab = 5$. Starting with $ab = 5$, you can see that the only ways to write 5 as a product of integers are $(1)(5) = 5$ and $(-1)(-5) = 5$. So, it is easy to see that you want $a = -1$ and $b = -5$ since $a + b = -1 + (-5) = -6$ and $ab = (-1)(-5) = 5$. This tells you that $x^2 - 6x + 5 = (x - 1)(x - 5)$.

For polynomials with four terms (usually a cubic polynomial), sometimes factoring by grouping works: You group the two higher-power terms and the two lower-power terms, factor the GCF out of each group, and then factor out the resulting common binomial factor, if there is one. For example, $x^3 + 5x^2 + 3x + 15 = (x^3 + 5x^2) + (3x + 15) = x^2(x + 5) + 3(x + 5) = (x^2 + 3)(x + 5)$.

Once you have found the factors, write the original polynomial as the product of all the factors. Make sure all of the factors are either monomials, or else linear or irreducible quadratic polynomials (*irreducible* means they have no real zeros, which is easy to check with the quadratic formula). Check your work by multiplying the factors to make sure you get the original polynomial.

Review Video: Factoring Out Common Monomial Factors
Visit mometrix.com/academy and enter code: 398578

Review Video: Factoring Trinomials of the Form x^2+bx+c
Visit mometrix.com/academy and enter code: 270556

Below are patterns of some special products to remember to help make factoring easier:

- Perfect square trinomials: $x^2 + 2xy + y^2 = (x + y)^2$ or $x^2 - 2xy + y^2 = (x - y)^2$. For example, $x^2 + 10x + 25 = (x + 5)^2$.
- Difference between two squares: $x^2 - y^2 = (x + y)(x - y)$. For example, $x^2 - 9 = (x + 3)(x - 3)$.
- Sum of two cubes: $x^3 + y^3 = (x + y)(x^2 - xy + y^2)$. For example, $x^3 + 27 = (x + 3)(x^2 - 3x + 9)$.
    - Note: the second factor is *not* the same as a perfect square trinomial, so do not try to factor it further.
- Difference between two cubes: $x^3 - y^3 = (x - y)(x^2 + xy + y^2)$. For example, $x^3 - 1000 = (x - 10)(x^2 + 10x + 100)$.
    - Again, the second factor is *not* the same as a perfect square trinomial.
- Perfect cubes: $x^3 + 3x^2y + 3xy^2 + y^3 = (x + y)^3$ and $x^3 - 3x^2y + 3xy^2 - y^3 = (x - y)^3$

Review Video: Factoring the Difference of Two Squares
Visit mometrix.com/academy and enter code: 128954

Rational and Irrational Expressions

Rational Expressions

Rational expressions are fractions with polynomials in both the numerator and the denominator; the value of the polynomial in the denominator cannot be equal to zero. Be sure to keep track of values that make the denominator of the original expression zero as the final result inherits the same restrictions. For example, a denominator of $x - 3$ indicates that the expression is not defined when $x = 3$ and, as such, regardless of any operations done to the expression, it remains undefined there.

To **add or subtract** rational expressions, first find the common denominator, then rewrite each fraction as an equivalent fraction with the common denominator. Finally, add or subtract the numerators to get the numerator of the answer, and keep the common denominator as the denominator of the answer.

When **multiplying** rational expressions, factor each polynomial and cancel like factors (a factor which appears in both the numerator and the denominator). Then, multiply all remaining factors in the numerator to get the numerator of the product, and multiply the remaining factors in the denominator to get the denominator of the product. Remember: cancel entire factors, not individual terms.

To **divide** rational expressions, take the reciprocal of the divisor (the rational expression you are dividing by) and multiply by the dividend.

Review Video: Rational Expressions
Visit mometrix.com/academy and enter code: 415183

Simplifying Rational Expressions

To simplify a rational expression, factor the numerator and denominator completely. Factors that are the same and appear in the numerator and denominator have a ratio of 1. For example, look at the following expression:

$$\frac{x-1}{1-x^2}$$

The denominator, $(1-x^2)$, is a difference of squares. It can be factored as $(1-x)(1+x)$. The factor $1-x$ and the numerator $x-1$ are opposites and have a ratio of -1. Rewrite the numerator as $-1(1-x)$. So, the rational expression can be simplified as follows:

$$\frac{x-1}{1-x^2} = \frac{-1(1-x)}{(1-x)(1+x)} = \frac{-1}{1+x}$$

Note that since the original expression is only defined for $x \neq \{-1, 1\}$, the simplified expression has the same restrictions.

Review Video: Reducing Rational Expressions
Visit mometrix.com/academy and enter code: 788868

Review Video: Simplifying Algebraic Expressions with Parentheses
Visit mometrix.com/academy and enter code: 850843

Irrational Expressions

Irrational expressions are mathematical expressions that contain an irrational number and cannot be simplified into a rational form. Usually, this includes expressions that contain radicals or constants such as π. Most commonly, you will encounter these in the forms of expressions containing roots of non-perfect squares.

Basic Operations on Radical Expressions

To add or subtract radical numbers, the numbers within the radicals must match, similar to finding a common denominator in fractions:

$$a\sqrt{x} + b\sqrt{x} = (a+b)\sqrt{x}$$

$$a\sqrt{x} - b\sqrt{x} = (a-b)\sqrt{x}$$

To multiply radicals, the numbers outside the radical are multiplied together and the numbers inside the radical are multiplied together:

$$a\sqrt{x} \times b\sqrt{y} = ab\sqrt{xy}$$

To divide radicals, the radical must be eliminated from the denominator by multiplying both numerator and denominator by a value that will make the denominator a rational number:

$$\frac{a\sqrt{x}}{b\sqrt{y}} = \frac{a\sqrt{x}\sqrt{y}}{b\sqrt{y}\sqrt{y}} = \frac{a\sqrt{xy}}{by}$$

Review Video: Adding and Subtracting Radical Expressions
Visit mometrix.com/academy and enter code: 752176

EXAMPLE

To solve $\frac{(3\sqrt{6})(2\sqrt{3})}{4\sqrt{5}}$, we first multiply the numerator, inside the radical and out: $3 \times 2\sqrt{6 \times 3} = 6\sqrt{18} = 18\sqrt{2}$. To divide, we multiply both numerator and denominator by a value that will eliminate the radical:

$$\frac{18\sqrt{2} \times \sqrt{5}}{4\sqrt{5} \times \sqrt{5}} = \frac{18\sqrt{10}}{4 \times 5} = \frac{18\sqrt{10}}{20} = \frac{9\sqrt{10}}{10}$$

Quadratics

SOLVING QUADRATIC EQUATIONS

A quadratic equation is an equation that can be written (possibly after simplification) in the form $ax^2 + bx + c = 0$. Thus, the **solutions** of this equation are precisely the **zeros** of the quadratic polynomial $P(x) = ax^2 + bx + c$. On the graph of this polynomial the zeros, if any, appear as x-intercepts. There are several ways to find these solutions including the quadratic formula, factoring, completing the square, and graphing the function.

Review Video: Quadratic Equations Overview
Visit mometrix.com/academy and enter code: 476276

Review Video: Solutions of a Quadratic Equation on a Graph
Visit mometrix.com/academy and enter code: 328231

QUADRATIC FORMULA

The **quadratic formula** gives the zeros of a quadratic polynomial. It always works, but it is sometimes a little harder to use than other methods. To use it to solve a quadratic equation, rewrite the equation in the form $ax^2 + bx + c = 0$, where a, b, and c are coefficients. Now, as explained above, the solutions of the equation are the zeros of the quadratic polynomial $P(x) = ax^2 + bx + c$. To find them, substitute the values of a, b, and c into the Quadratic Formula:

$$x = \frac{-b \pm \sqrt{b^2 - 4ac}}{2a}$$

After simplification this formula produces two, one, or zero real solutions, depending on whether the **discriminant** (the number $b^2 - 4ac$ under the radical) is positive, zero, or negative. It is a good practice to check each solution by substituting it into the original equation. Incidentally, if the discriminant is negative, then the equation does have two complex solutions, but you often ignore these as meaningless in real-world settings.

Review Video: Using the Quadratic Formula
Visit mometrix.com/academy and enter code: 163102

FACTORING

To solve a quadratic equation by factoring, begin by rewriting the equation in the standard form, $ax^2 + bx + c = 0$. In the important special case that $a = 1$, the goal of factoring is to find numbers f and g such that $x^2 + bx + c = (x + f)(x + g) = x^2 + (f + g)x + fg$. In other words, you want to choose f and g to make $fg = c$ and $f + g = b$. To do this, find pairs of numbers (factors) whose product is c and look for a pair whose sum is b.

For example, suppose you want to find the solutions of the equation $x^2 + 6x - 16 = 0$ by factoring. Here $b = 6$ and $c = -16$. First, you find the pairs of numbers whose product is -16. These are -4 and 4, -8 and 2, -2 and 8, -1 and 16, and 1 and -16. The pair -2 and 8 has a sum of 6. This means $f = -2$ and $g = 8$. So, the factorization is $x^2 + 6x - 16 = (x + f)(x + g) = (x - 2)(x + 8)$, allowing you to rewrite the original equation as $(x - 2)(x + 8) = 0$. The only way for a product to equal zero is for one of the factors to equal zero; so, either $x - 2 = 0$ (in which case $x = 2$) or $x + 8 = 0$ (in which case $x = -8$). Thus, the equation has the solution $x = 2$ or $x = -8$.

In the case that $a \neq 1$, you can attempt to factor the quadratic polynomial $ax^2 + bx + c$ in the form $(mx + f)(nx + g)$ by a similar trial-and-error procedure, but the work tends to be much harder.

Review Video: Factoring Quadratic Equations
Visit mometrix.com/academy and enter code: 336566

COMPLETING THE SQUARE

The technique of completing the square comes from a simple observation: Suppose you have the expression $x^2 + bx$. If you take half the linear coefficient, b, square it, and add the result to the expression, the result is always a perfect square trinomial:

$$x^2 + bx + \left(\frac{b}{2}\right)^2 = \left(x + \frac{b}{2}\right)^2$$

For example, if you begin with $x^2 + 6x$ and add the square of half of 6 (half of 6 is 3, and $3^2 = 9$), then you get a perfect square trinomial:

$$x^2 + 6x + 9 = (x + 3)^2$$

This also works if b is negative. For instance, if you begin with the expression $x^2 - 10x$ and complete the square by adding 25 (half of -10 is -5, and $(-5)^2 = 25$), then you get another perfect square trinomial:

$$x^2 - 10x + 25 = (x - 5)^2$$

Now, suppose you want to solve the equation $x^2 + bx + c = 0$. Subtract c from both sides, to get the equation $x^2 + bx = -c$. Complete the square on the left side by adding $(b/2)^2$, but also add this same term to the right side so that the new equation is equivalent to the old one:

$$x^2 + bx + \left(\frac{b}{2}\right)^2 = -c + \left(\frac{b}{2}\right)^2$$
$$\left(x + \frac{b}{2}\right)^2 = -c + \left(\frac{b}{2}\right)^2$$

Take square roots of both sides:

$$x + \frac{b}{2} = \pm\sqrt{-c + \left(\frac{b}{2}\right)^2}$$

Remember to include $\pm$ since every positive number has both a positive and a negative square root. Subtract $b/2$ from both sides to isolate the variable x to finish the problem:

$$x = -\frac{b}{2} \pm \sqrt{-c + \left(\frac{b}{2}\right)^2}$$

This may sound complicated, but in practice it is not hard. For example, suppose you want to solve the equation $x^2 + 6x - 16 = 0$ by completing the square. First, add 16 to both sides:

$$x^2 + 6x = 16$$

Half of 6 is 3 and $3^2 = 9$, so complete the square by adding 9 to both sides of the equation:

$$x^2 + 6x + 9 = 16 + 9$$
$$(x + 3)^2 = 25$$

Now take square root of both sides, remembering to include the $\pm$:

$$\sqrt{(x + 3)^2} = \pm\sqrt{25}$$
$$x + 3 = \pm 5$$
$$x = -3 \pm 5$$

So we see that the two solutions to the equation are $x = 2$ and $x = -8$.

Review Video: Completing the Square
Visit mometrix.com/academy and enter code: 982479

Using Given Solutions to Find a Quadratic Equation

To find a quadratic equation with given numbers as solutions, simply find a quadratic polynomial that has those numbers as zeros and set that polynomial equal to zero. This is easy because a polynomial has the number p as a zero precisely when it has the binomial $x - p$ as a factor. Thus, for instance, to find a quadratic polynomial with zeros at $x = 3$ and $x = -5$, construct the polynomial $P(x) = (x - 3)\big(x - (-5)\big) = (x - 3)(x + 5) = x^2 + 2x - 15$. Setting this equal to zero produces an equation, $x^2 + 2x - 15 = 0$, whose solutions are $x = 3$ and $x = -5$.

Of course, any constant multiple $P(x) = a(x-3)(x+5)$ will also have the same zeros. For instance, if you choose $a = 4$, then you get the polynomial $P(x) = 4(x-3)(x+5) = 4x^2 + 8x - 60$, which also has zeros at $x = 3$ and $x = -5$. From this you can get another equation, $4x^2 + 8x - 60 = 0$, with solutions $x = 3$ and $x = -5$.

Chapter Quiz

Ready to see how well you retained what you just read? Scan the QR code to go directly to the chapter quiz interface for this study guide. If you're using a computer, simply visit the online resources page at **mometrix.com/resources719/cchsalgebra** and click the Chapter Quizzes link.

Functions

Algebraic Theorems

According to the **fundamental theorem of algebra**, every non-constant, single-variable polynomial has exactly as many roots as the polynomial's highest exponent. For example, if x^4 is the largest exponent of a term, the polynomial will have exactly 4 roots. However, some of these roots may have multiplicity or be complex numbers. For instance, in the polynomial function $f(x) = x^4 - 4x + 3$, the only real root is 1, though it has multiplicity of 2 – that is, it occurs twice. The other two roots, $(-1 - i\sqrt{2})$ and $(-1 + i\sqrt{2})$, are complex, consisting of both real and non-real components.

The **remainder theorem** is useful for determining the remainder when a polynomial is divided by a binomial. The remainder theorem states that if a polynomial function $f(x)$ is divided by a binomial $x - a$, where a is a real number, the remainder of the division will be the value of $f(a)$. If $f(a) = 0$, then a is a root of the polynomial.

The **factor theorem** is related to the remainder theorem and states that if $f(a) = 0$ then $(x - a)$ is a factor of the function.

According to the **rational root theorem,** any rational root of a polynomial function $f(x) = a_n x^n + a_{n-1}x^{n-1} + \cdots + a_1 x + a_0$ with integer coefficients will, when reduced to its lowest terms, be a positive or negative fraction such that the numerator is a factor of a_0 and the denominator is a factor of a_n. For instance, if the polynomial function $f(x) = x^3 + 3x^2 - 4$ has any rational roots, the numerators of those roots can only be factors of 4 (1, 2, 4), and the denominators can only be factors of 1 (1). The function in this example has roots of 1 (or $\frac{1}{1}$) and -2 (or $\frac{-2}{1}$).

Parabolas

A **parabola** is the set of all points in a plane that are equidistant from a fixed line, called the **directrix**, and a fixed point not on the line, called the **focus**. The **axis** is the line perpendicular to the directrix that passes through the focus.

For parabolas that open up or down, the standard equation is $(x - h)^2 = 4c(y - k)$, where h, c, and k are coefficients. If c is positive, the parabola opens up. If c is negative, the parabola opens down. The vertex is the point (h, k). The directrix is the line having the equation $y = -c + k$, and the focus is the point $(h, c + k)$.

For parabolas that open left or right, the standard equation is $(y - k)^2 = 4c(x - h)$, where k, c, and h are coefficients. If c is positive, the parabola opens to the right. If c is negative, the parabola opens to the left. The vertex is the point (h, k). The directrix is the line having the equation $x = -c + h$, and the focus is the point $(c + h, k)$.

> **Review Video: Parabolas**
> Visit mometrix.com/academy and enter code: 129187
>
> **Review Video: Vertex of a Parabola**
> Visit mometrix.com/academy and enter code: 272300
>
> **Review Video: How to Convert Between Standard Form and Vertex Form**
> Visit mometrix.com/academy and enter code: 916601

Basics of Functions

Definition of a Function

A function is a rule that assigns to every number in a given set (called the **domain**) exactly one corresponding value. For example, if our domain is the set $\{-2,1,2,3\}$, we can define a function by assigning to each number its square. This function assigns to -2 the value 4, to 1 the value 1, to 2 the value 4, and to 3 the value 9 (since $(-2)^2 = 4$, $1^2 = 1$, $2^2 = 4$, and $3^2 = 9$). The set of all the values assigned by a function is the **range** of the function. The range of the function in our example is the set $\{1, 4, 9\}$. We may think of a function as a kind of machine: we give it a number as an input, and it uses its rule to produce a number as an output. In the squaring function above, the input 3 produces the output 9.

> **Review Video: What is a Function?**
> Visit mometrix.com/academy and enter code: 784611

Function Notation

We usually name a function by a letter, often the letter f (for *function*—if we need to talk about more than one function, we name the second one g, the third one h, etc.). To specify the value (the output) corresponding to a particular number in the domain (the input), we write the function letter followed by the input number in parentheses. For instance, in the example above the notation $f(3)$ means the value that the function assigns to the number 3, namely 9—that is, $f(3) = 9$. We read the symbols $f(3)$ as, "f of 3," and we call 3 the **argument** of the function and 9 the **value** of the function (so *argument* means *input* and *value* means *output*).

Using function notation we can define the squaring function above by listing the values the function assigns to each argument in the domain: $f(-2) = 4$. $f(1) = 1$, $f(2) = 4$, and $f(3) = 9$. More efficiently, we can define the function by the single equation $f(x) = x^2$, which says that if x is a number from the domain, then we calculate the value assigned to it by substituting the number x in the formula x^2. For instance, we calculate $f(5) = 5^2 = 25$. Similarly, if we define a function g by the equation $g(x) = x^2 - 4x + 7$, then we calculate the value $g(3)$ by substituting 3 for each x in the formula: $g(3) = 3^2 - 4 \cdot 3 + 7 = 9 - 12 + 7 = 4$.

Other Ways to Define Functions

Instead of denoting the value of the function by $f(x)$, sometimes we simply use another letter, usually y. For instance, instead of defining the squaring function by the equation $f(x) = x^2$, we might use the equation $y = x^2$. In this case, we refer to x (the input) as the **independent variable** and y (the output) as the **dependent variable** because the value, y, depends on the number we choose for x.

A formula (with y or $f(x)$) is the most common way to define a function; but sometimes, if the domain is small enough, we prefer to list explicitly the possible inputs and their corresponding outputs. Some ways of doing this appear above, but a more common approach is to put the input-output pairs in a table. For instance, we can define the squaring function above by the table

x	−2	1	2	3
y	4	1	4	9

We see that the domain of this function is the set of all numbers in the x-row and the range is the set of all numbers in the y-row. We note that numbers cannot repeat in the x-row (because a function assigns exactly one value to each argument in the domain) but they can repeat in the y-row (because the function can assign the same value to multiple arguments—for instance, the number 4 appears twice in the y-row).

We can also define a function by writing the inputs and corresponding outputs as ordered pairs of x- and y-values. For instance, we can write the squaring function above as the set of ordered pairs $\{(-2,4), (1,1), (2,4), (3,9)\}$. Further, by treating these ordered pairs as coordinates and plotting the corresponding points on the coordinate plane, we get the **graph** of the function:

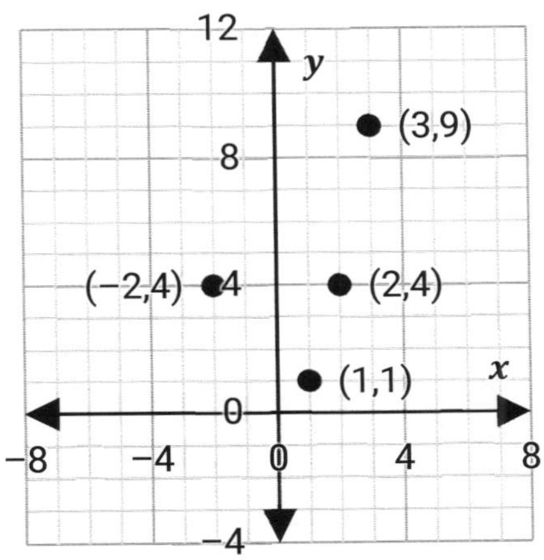

Turning this around, we can potentially use a graph to define a function, namely the function consisting of the coordinate pairs of all the points in the graph. This always works unless the graph has two points with the same x-coordinate (because then the function would assign two different y-values to the same x). It is easy to detect such points: because they have the same x-coordinate, a vertical line passes through both of them. Thus, a graph always defines a function unless it is possible to draw a vertical line that intersects the graph in two or more points. We call this condition the **vertical line test**. For example, if our graph is a circle, then by the Vertical Line Test the graph does not define a function because there are vertical lines that will intersect the circle in two different points.

More on Domains and Ranges

When we define a function by a formula and do not specify the domain, then by default the domain consists of all real numbers for which the formula produces an answer. For instance, suppose we define a function f by the formula $f(x) = 1/x$. If $x = 0$, then $1/x = 1/0$, which is undefined. But if x is any other real number, then we can calculate the value of $1/x$. So, the default domain of this function is all real numbers except zero. Because of this domain convention, the graph of a function defined by a formula usually consists of infinitely many points that "connect to" each other in a way that produces a line or curve (see examples below) rather than the isolated points we see in the squaring function above.

If we have the graph of a function, its domain consists of all numbers on the x-axis with corresponding points on the graph and its range consists of all numbers on the y-axis with corresponding points on the graph. For example, consider the function $f(x) = x^2 + 3$:

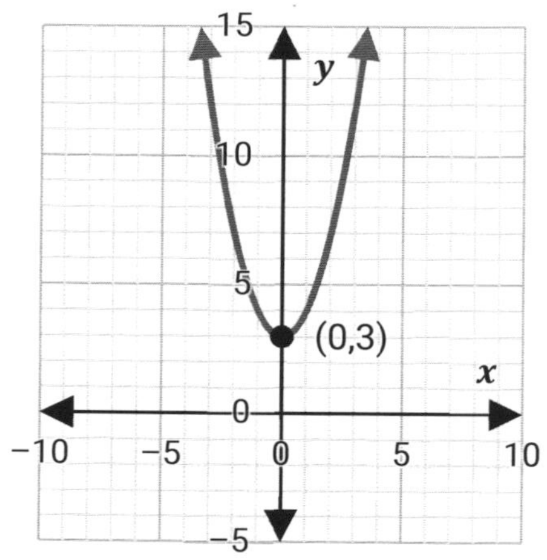

Since the graph continues infinitely to the left and right beyond what we can see, every point on the x-axis has a corresponding point on the graph; so, the domain of this function is all real numbers. On the other hand, the lowest point on this graph has a y-value of 3, and the graph passes through all higher y-values. So, the range of this function is all real numbers greater than or equal to 3, which we can denote algebraically by $y \geq 3$ or, using interval notation, by $[3, \infty)$.

> **Review Video: How to Find Domain and Range**
> Visit mometrix.com/academy and enter code: 778133
>
> **Review Video: Domain and Range of Quadratic Functions**
> Visit mometrix.com/academy and enter code: 331768

Monotonic and Even/Odd Functions

A function, f, is **increasing** if it always assigns larger values to larger arguments. It is **decreasing** if it always assigns smaller values to larger arguments. That is, f is increasing if $a < b$ always guarantees $f(a) < f(b)$, and it is decreasing if $a < b$ always guarantees $f(a) > f(b)$. The graph of an increasing function consistently rises from left to right, and the graph of a decreasing function consistently falls from left to right. For example, the function $f(x) = 2x$ is an increasing function because doubling a larger number always gives us a larger result than doubling a smaller number.

The graph of $f(x) = 2x$ is a line with slope $m = 2$, which, as we expect, rises from left to right. We call a function **monotonic** if it is either increasing or decreasing.

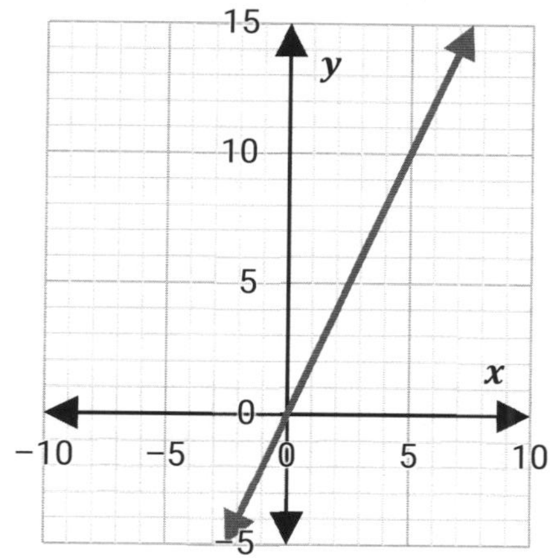

A function, f, is **even** if changing the sign of its argument produces the same value. It is **odd** if changing the sign of its argument produces the same value except with the opposite sign. That is, f is even if $f(-x) = f(x)$ and odd if $f(-x) = -f(x)$ for every argument x. The function $f(x) = x^2 + 3$ is even because substituting opposite arguments always produces the same value. For instance, $f(5) = 28$ and $f(-5) = 28$ because $5^2 + 3 = 25 + 3 = 28$ and $(-5)^2 + 3 = 25 + 3 = 28$. The function $f(x) = 2x$ is odd because substituting opposite arguments always produces opposite values. For instance, $f(10) = 20$ and $f(-10) = -20$ because $2(10) = 20$ and $2(-10) = -20$. The graph of an even function is always symmetric with respect to the y-axis, making the left and right halves of the graph mirror images of each other, as in the graph of the even function $f(x) = x^2 + 3$ above. The graph of an odd function is always symmetric with respect to the origin. This means that if we rotate the graph 180° around the origin (think of sticking a pin through the origin on a sheet of graph paper and rotating the paper halfway around) the graph looks the same, as in the graph of the odd function $f(x) = 2x$ above.

It is worth noting that most functions are neither increasing nor decreasing (that is, they are not monotonic) and most functions are neither even nor odd. For example, the function $f(x) = x^2 -$

x is neither increasing nor decreasing and neither even nor odd: its graph neither rises nor falls consistently, and it is symmetric with respect to neither the y-axis nor the origin.

Review Video: Even and Odd Functions
Visit mometrix.com/academy and enter code: 278985

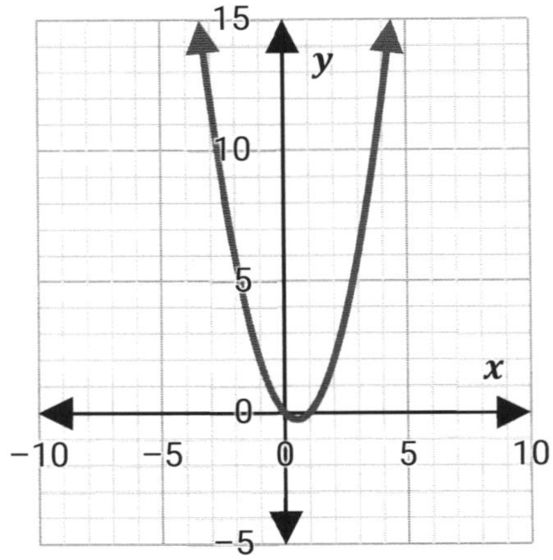

INVERTIBLE (ONE-TO-ONE) FUNCTIONS

A function, f, is one-to-one if it never assigns the same value to different arguments—that is, if $f(a)$ and $f(b)$ are different whenever a and b are different. The graph of a one-to-one function never has two points that lie on the same horizontal line because such points would have different x-values but the same y-value. Thus, a function is one-to-one if it is impossible to draw a horizontal line that intersects its graph in more than one point. We call this condition the **horizonal line test**. For example, the graph of the function $f(x) = 2x$ above is a line that rises from left to right. Every horizontal line intersects this line in exactly one point, so the function $f(x) = 2x$ is one-to-one. This is also clear without the graph because it is impossible to double two different numbers and get the same answer.

When a function, f, is one-to-one, it is possible to define its inverse function, f^{-1}, that "undoes" what f does, assigning to each output from f the input that produced it. That is, for each x in the domain of f, if $y = f(x)$, then $f^{-1}(y) = x$. For example, the inverse of the function $f(x) = 2x$ above is $f^{-1}(y) = y/2$. So, for instance, $f(5) = 2 \cdot 5 = 10$, and $f^{-1}(10) = 10/2 = 5$ (and similarly for every other value of x). Thus, the domain of f^{-1} is the range of f and vice versa. If a function, f, has an inverse, we say that f is **invertible**. Since a function has an inverse precisely when it is one-to-one, the terms *invertible* and *one-to-one* are synonyms.

If f is an invertible function defined by a formula, then to find its inverse we simply write the equation $y = f(x)$ and solve it for x (that is, we isolate the x). The result will be the equation $f^{-1}(y) = x$. For instance, starting with the function $f(x) = 2x$, we write $y = 2x$ and isolate the x by dividing both sides of the equation by 2. This gives us $y/2 = x$, so we know that $f^{-1}(y) = y/2$. Although this procedure is theoretically simple, in practice the algebra can be difficult.

Common Functions

Certain functions and certain kinds of functions are particularly useful, coming up frequently in mathematics and its applications. Once we know some basic function terminology and concepts, it is useful to begin developing a mental library of the most common and useful functions.

Review Video: Common Functions
Visit mometrix.com/academy and enter code: 629798

Constant Functions

A function of the form $f(x) = a$, where a is a real number, is a **constant function**. This function assigns the same value, a, to every real argument x. For instance, given the constant function $f(x) = 5$, we have $f(2) = 5$, $f(100) = 5$, and $f(-7.1) = 5$.

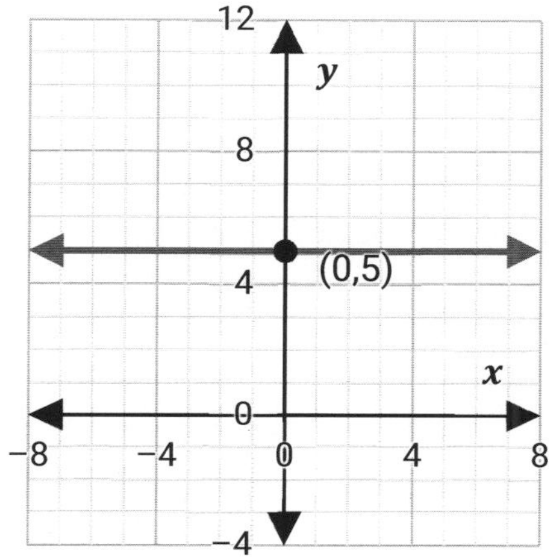

The domain of a constant function is the set of all real numbers, and the range is the set containing the single number a. Its graph is a horizontal line passing through the number $y = a$ on the y-axis (we call the number at which a function's graph intersects the y-axis the ***y*-intercept** of the function).

THE IDENTITY FUNCTION

The function $f(x) = x$ is the **identity function**. Its value always equals its argument. Thus, for instance, $f(2) = 2$, $f(100) = 100$, and $f(-7.1) = -7.1$.

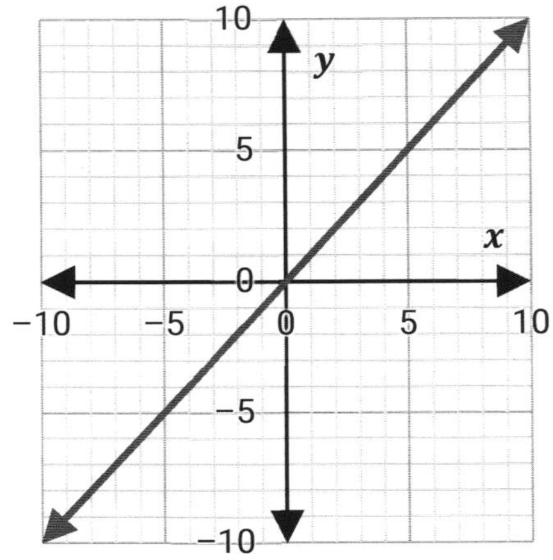

Its domain and range are the set of all real numbers. It is both an increasing function and an odd function. Its graph is a line that passes through the origin and rises from left to right at a 45° angle to the horizontal. Since it passes through the origin, its y intercept is $y = 0$ and it also has an ***x*-intercept** (a number at which the function's graph intersects the x-axis) of $x = 0$.

LINEAR FUNCTIONS

A function of the form $f(x) = ax + b$, where a and b are real numbers (with $a \neq 0$), is a **linear function** (the identity function is a linear function with $a = 1$ and $b = 0$). Its domain and range are the set of all real numbers. Its graph is a line (the word *linear* contains the root word *line*) with one x-intercept (at $x = -b/a$), with a y-intercept at $y = b$, and with a direction and steepness that depend on the coefficient a, which we call the **slope**. Specifically, the slope a is the amount the y-value increases for each increase of 1 in the x-value. Thus, for $a > 0$, the line rises from left to right (making f an increasing function), and larger values of a produce steeper ascents. Similarly, for $a < 0$, the line falls from left to right (making f a decreasing function), and smaller (more negative) values of a produce steeper descents. For instance, the graph of the linear function

$f(x) = (1/2)x + 3$ is a line that passes through the point $y = 3$ on the y-axis and that rises by $1/2$ unit for every unit that x increases.

Review Video: Linear Functions
Visit mometrix.com/academy and enter code: 200735

Review Video: Graphing Linear Functions
Visit mometrix.com/academy and enter code: 699478

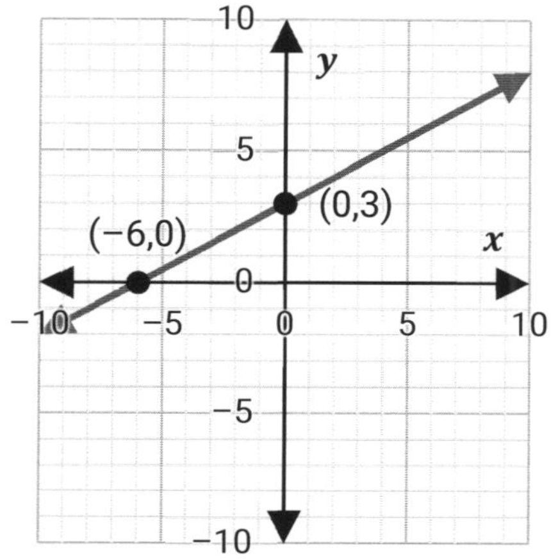

In many contexts it is standard to use the letter m for slope and thus to write the general form of a linear function as $f(x) = mx + b$, known as **slope-intercept form.**

THE SQUARING FUNCTION

The function $f(x) = x^2$ is the **squaring function**.

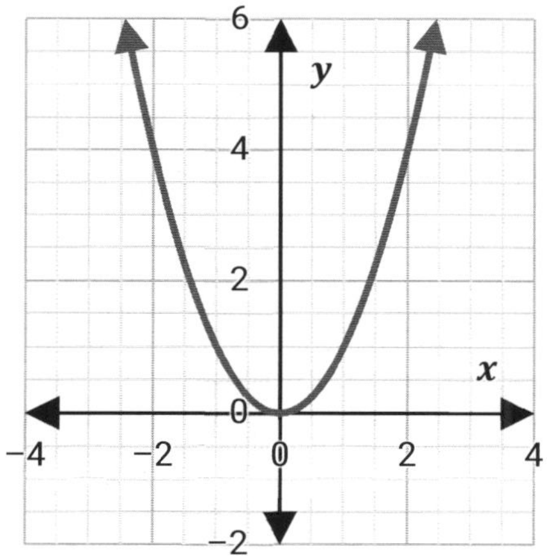

Its graph is U-shaped, opening upward as shown, a shape known as a **parabola**. It has a lowest point, its **vertex**, at the origin, which is also the location of its single x-intercept and single y-intercept. Thus, its **minimum** is $y = 0$, its domain is the set of all real numbers, and its range is the set of nonnegative real numbers (that is, $y \geq 0$). It is an even function and thus symmetric with respect to the y-axis (which we call the **axis of symmetry**), meaning that the left half of the graph is the mirror image of the right half, with the mirror standing on the y-axis.

QUADRATIC FUNCTIONS

A function of the form $f(x) = ax^2 + bx + c$, where a, b, and c are real numbers (with $a \neq 0$), is a **quadratic function** (the squaring function is a quadratic function with $a = 1$, $b = 0$, and $c = 0$). Its domain is the set of all real numbers, and its graph is a parabola. It is symmetric with respect to its axis of symmetry, the vertical line $x = -b/(2a)$. If $a > 0$, the parabola opens upward, so that its vertex is at its lowest point (its minimum) and its range consists of all real numbers greater than or equal to this minimum y-value. If $a < 0$, the parabola opens downward, so that its vertex is at its highest point (its maximum) and its range consists of all real numbers less than or equal to this maximum y-value. Its y-intercept is $y = c$ since $f(0) = c$, and it may have zero, one, or two x-intercepts. For example, the function $f(x) = x^2 - 6x + 5$ has $a = 1$, $b = -6$, and $c = 5$.

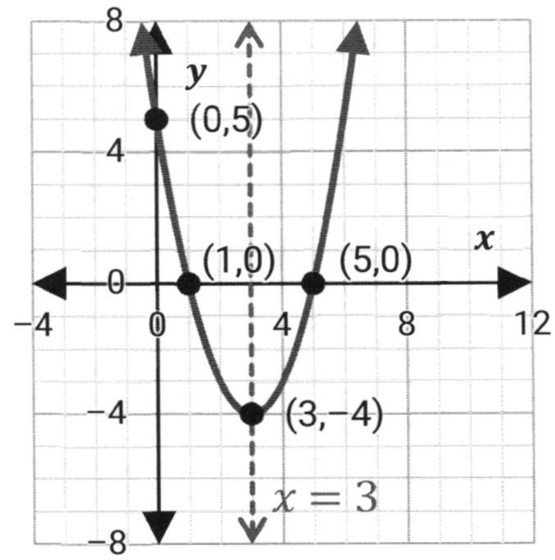

Its graph opens upward (because $a > 0$) and its axis of symmetry is the vertical line $x = 3$ (since $-b/(2a) = -(-6)/(2 \cdot 1) = 3$). Its y-intercept is at $y = 5$. It turns out to have its vertex at the point $(3, -4)$, making its minimum value $y = -4$. So, its domain is the set of all real numbers, and its range is $y \geq -4$. It also turns out to have two x-intercepts, at $x = 1$ and at $x = 5$ (since $f(1) = 0$ and $f(5) = 0$).

POLYNOMIAL FUNCTIONS

A function of the form $f(x) = a^n x^n + a^{n-1} x^{n-1} + \cdots + a_2 x^2 + a_1 x + a_0$, where n is a whole number and $a_0, a_1, a_2, \ldots a_{n-1}, a_n$ are real numbers, is a **polynomial function of degree n**. Its domain is the set of all real numbers (it is complicated to describe its range in general), and its y-intercept is $y = a_0$ (since $f(0) = a_0$). Constant functions, linear functions, and quadratic functions are polynomial functions of degrees 0, 1, and 2, respectively. In general, a polynomial function of degree n has up to n zeros (x-intercepts) and up to $n - 1$ "bends." For example, the fourth degree polynomial function $f(x) = x^4 - 11x^3 + 41x^2 - 61x + 30$, whose graph appears here, has four x-intercepts (at $x = 1, x = 2, x = 3$, and $x = 5$) and three "bends," and its y-intercept (not visible on the graph) is at $y = 30$.

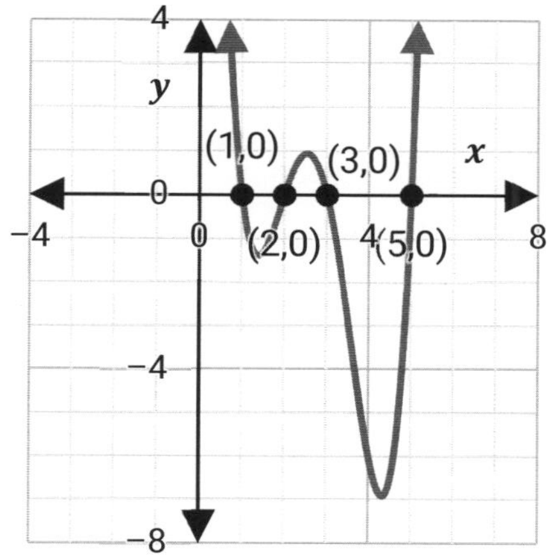

Rational Functions

A function of the form $f(x) = P(x)/Q(x)$, where P and Q are polynomials, is a rational function (we note that the word *rational* includes the root word *ratio*, indicating that a rational function is a ratio of polynomial functions). The domain of a rational function is all real numbers except the zeros of $Q(x)$ since division by zero is undefined (the range can be difficult to describe in general). Its y-intercept is $f(0)$, if this is defined; and its x-intercepts are the zeros of $P(x)$ that are in the domain of f, if there are any. A rational function may also have vertical asymptotes (vertical lines that the graph approaches without crossing) and a horizontal asymptote (a horizontal line that the curve approaches as x becomes very small or very large (toward the left and right edges of the graph). For example, the rational function $f(x) = (2x^2 + x - 1)/(x^2 + x - 2)$ has as its domain the set of all real numbers except $x = -2$ and $x = 1$ (since those numbers make the denominator zero).

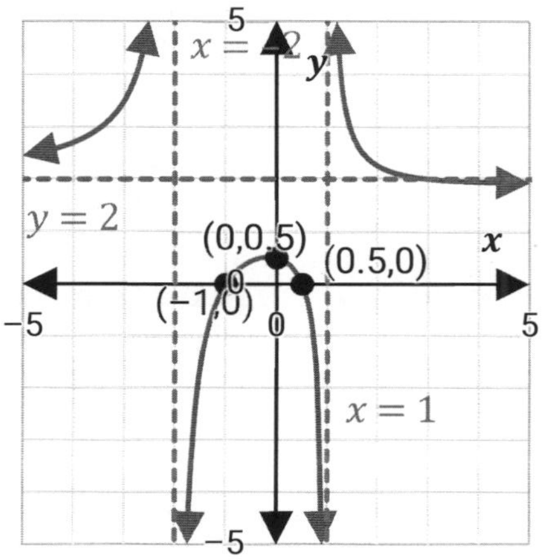

It has a y-intercept of $y = 1/2$ since $f(0) = (-1)/(-2) = 1/2$, and it has x-intercepts at $x = -1$ and at $x = 1/2$ since those numbers make the numerator zero. It has vertical asymptotes at $x = -2$ and $x = 1$ (not coincidentally, these are the numbers omitted from the domain) and a horizontal asymptote at $y = 2$. It is important to note that vertical asymptotes cannot be crossed in rational functions, but horizontal asymptotes can be crossed if the function tends near the asymptote at infinity and does not go past all possible turning points.

> **Review Video: Simplifying Rational Polynomial Functions**
> Visit mometrix.com/academy and enter code: 351038
>
> **Review Video: Horizontal Asymptotes**
> Visit mometrix.com/academy and enter code: 747796

The Square Root Function

The function $f(x) = \sqrt{x}$ is the square root function.

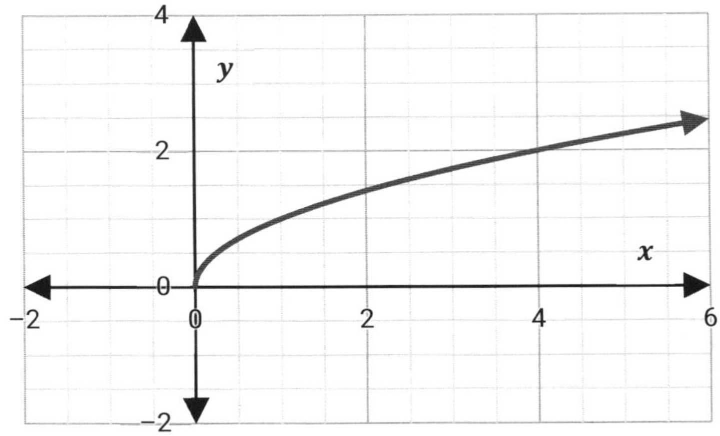

It is an increasing function, and its domain and range are both the set of all nonnegative real numbers. It has one x-intercept and one y-intercept, both appearing at the origin. Its graph is the upper half of a parabola opening to the right. The square root function is the inverse of the squaring function with domain restricted to the nonnegative real numbers (that is, $f(x) = x^2$ for $x \geq 0$).

Piecewise-Defined Functions

As the name suggests, a **piecewise-defined function** (or, simply, a **piecewise function**) is a function defined by different rules on different pieces of the domain. We define such a function using the following form:

Function Name	Rule to Apply	Piece of the Domain on Which the Rule Applies
$f(x) =$	Rule 1,	First Piece of the Domain
	Rule 2,	Second Piece of the Domain
	Rule 3,	Third Piece of the Domain
	etc.,	etc.

The pieces of the domain should not overlap, and together they should cover the whole domain. For example, we might craft a piecewise-defined function by

$$f(x) = \begin{cases} x^2, & \text{if } x < 2 \\ 3x - 5, & \text{if } x \geq 2 \end{cases}$$

The two pieces of the domain—namely, $x < 2$ and $x \geq 2$—do not overlap, and together they include all real numbers. To evaluate the function for a particular argument x, we determine which piece of the domain includes x and then apply the corresponding rule. For instance, to find $f(4)$, we note that $4 \geq 2$; so, we apply the rule $3x - 5$ to get the value $f(4) = 3 \cdot 4 - 5 = 7$. Similarly, to find $f(-6)$, we note that $-6 < 2$; so, we apply the rule x^2 to get the value $f(-6) = (-6)^2 = 36$.

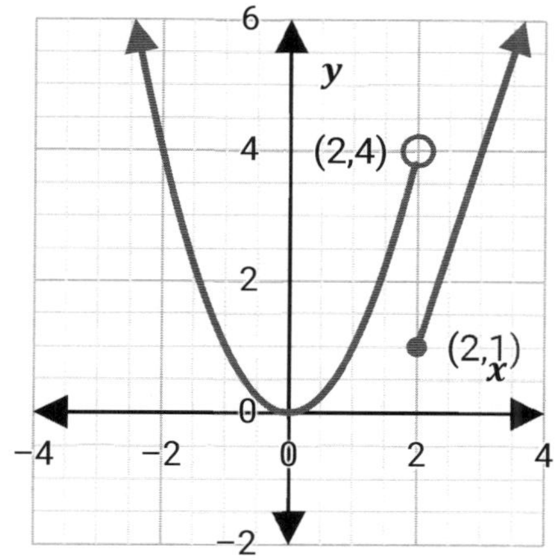

To graph this function, we sketch the graph of the parabola $y = x^2$ on the part of the plane where $x < 2$ and we sketch the line $y = 3x - 5$ on the part of the plane where $x \geq 2$. This produces a graph with a jump at $x = 2$ (a discontinuity—piecewise-defined functions are useful for producing graphs with discontinuities). We plot an open circle at the point (2,4), the end of the left part of the graph, to show that this point is not part of the graph. And we plot a solid dot at the point (2,1), the start of the right part of the graph, to show that this point *is* part of the graph.

Review Video: Piecewise Functions
Visit mometrix.com/academy and enter code: 707921

The Absolute Value Function

A particularly useful piecewise-defined function is the absolute value function. It is so important that instead of naming it $f(x)$ or $g(x)$, we denote it using the special notation $|x|$. Its definition is

$$|x| = \begin{cases} -x, & \text{if } x < 0 \\ x, & \text{if } x \geq 0 \end{cases}$$

For instance, $|8| = 8$ (since $8 \geq 0$) and $|-5| = -(-5) = 5$, since $-5 < 0$. So, the absolute value function acts like the identity function for nonnegative numbers (it leaves them unchanged), and it gives the opposite of negative numbers (it effectively strips off the minus sign). Thus, we can think of the absolute value of a real number as its distance from zero on the number line, without taking into consideration whether the number is larger than or smaller than zero. For instance, $|-3| = 3$ and $|3| = 3$, showing that both -3 and 3 are three units away from zero. The absolute value function is an even function with a V-shaped graph that looks like the line $y = x$ (the identity function) on the right "half" of the plane (for $x \geq 0$) and the line $y = -x$ on the left "half" of the plane (for $x < 0$).

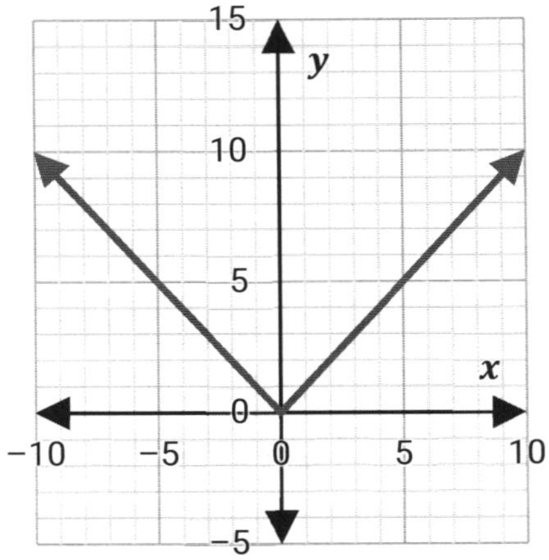

Trigonometric Functions

The basic trigonometric functions are sine (abbreviated 'sin'), cosine (abbreviated 'cos'), and tangent (abbreviated 'tan'). The simplest way to think of them is as describing the ratio of the side lengths of a right triangle in relation to the angles of the triangle.

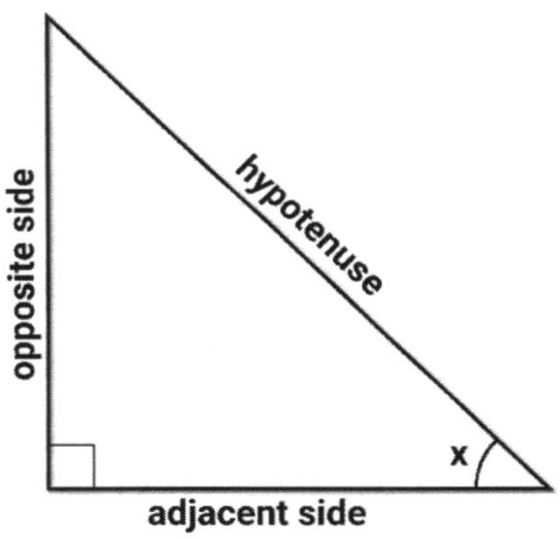

$$\sin x = \frac{\text{opposite side}}{\text{hypotenuse}}$$
$$\cos x = \frac{\text{adjacent side}}{\text{hypotenuse}}$$
$$\tan x = \frac{\text{opposite side}}{\text{adjacent side}}$$

Applying Trigonometric Functions

Suppose we have a point (x, y) at distance r from the origin such that the line segment between this point and the origin forms an angle of measure θ with the positive x-axis (We allow θ to be any real number, with the angle measured counterclockwise from the x-axis if $\theta > 0$ and clockwise if $\theta < 0$). Then we can define the trigonometric functions sine, cosine, and tangent by the formulas $\sin\theta = y/r$, $\cos\theta = x/r$, and $\tan\theta = y/x$ (and their reciprocal functions cosecant, secant, and cotangent accordingly). These functions are periodic because increasing the angle θ by 2π radians (that is,

360°) turns the line segment all the way around the origin so that it ends at the same point (x, y) again.

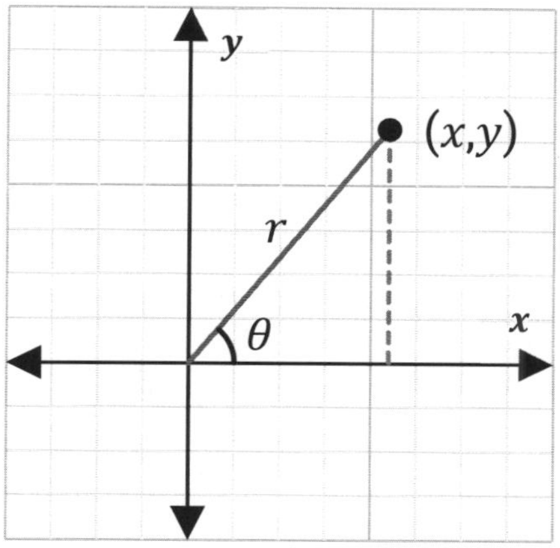

If the angle in the above diagram is acute, we can think of the line segment as the hypotenuse of a right triangle so that we get the familiar formulas $\sin\theta = (\text{opposite leg})/\text{hypotenuse}$, $\cos\theta = (\text{adjacent leg})/\text{hypotenuse}$, and $\tan\theta = (\text{opposite leg})/(\text{adjacent leg})$. In this form, the trigonometric functions are useful for finding missing sides of triangles. For example, suppose that standing 30 feet from the base of a tree, we have to look up at a 58° angle to see the top of the tree (see diagram). If we let x be the height of the tree, then we get the equation $\tan 58° = x/(30\text{ feet})$. Multiplying both sides of the equation by 30 gives us the height of the tree, namely, $x = 30\tan 58° \approx 48$ feet.

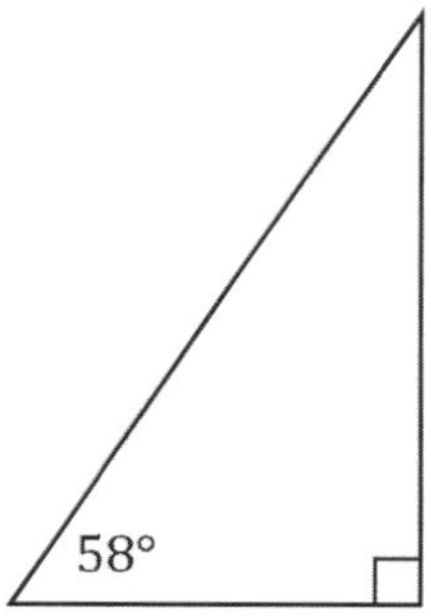

Graphing Trigonometric Functions

Sine

To graph the sine function, we let x (rather than θ) be the angle measure (usually in radians) and y be the value of the sine. That is, we graph the equation $y = \sin x$. From the definition of sine above we can show that $\sin 0 = 0$, $\sin(\pi/2) = 1$, $\sin\pi = 0$, $\sin(3\pi/2) = -1$, and $\sin 2\pi = 0$. Thus, the graph of the sine function begins at the origin, peaks at the point $(\pi/2, 1)$, crosses the x-axis at $(\pi, 0)$, has its lowest point at $(3\pi/2, -1)$, and returns to the x-axis at $(2\pi, 0)$. Because the sine function is periodic with period 2π radians (or 360°), this curve represents one complete cycle of the graph. The remainder of the graph of $y = \sin x$ consists of infinitely many repetitions of this cycle, producing the familiar "sine wave."

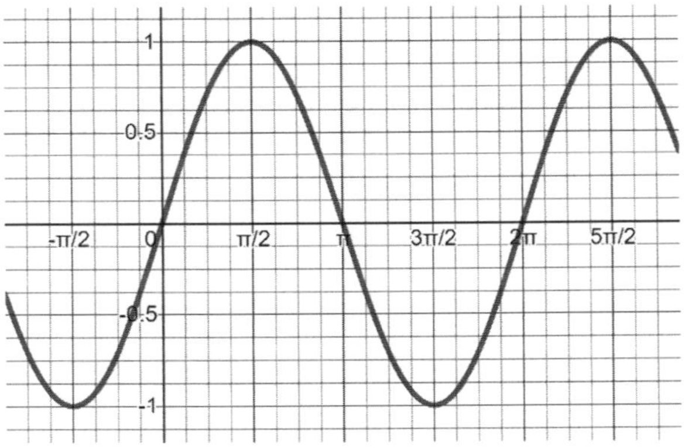

Review Video: Sine
Visit mometrix.com/academy and enter code: 193339

By suitable choice of the parameters A, B, C, and D, we can construct a "custom" sine function $y = A\sin(Bx + C) + D$ with any desired amplitude (the vertical distance the curve extends above and below its midline), period (distance between successive peaks), phase shift (horizontal shift), and midline (the horizontal line halfway between the peaks and troughs).

Cosine

Similarly, the graph of the cosine function is the graph of the equation $y = \cos x$. Proceeding as above, we get a graph with the same shape as that of the sine function but shifted by $\pi/2$ to the left.

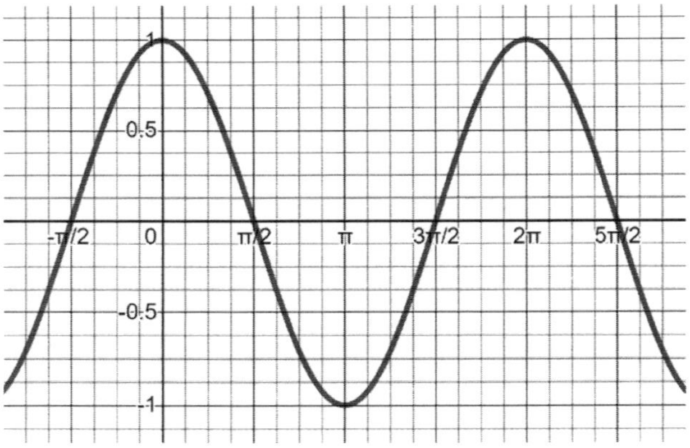

Review Video: Cosine
Visit mometrix.com/academy and enter code: 361120

As for the sine function, we can construct custom cosine graphs with desired amplitude, period, phase shift, and midline by choosing suitable values of the parameters in the equation $y = A\cos(Bx + C) + D$.

TANGENT

The graph of the tangent function (that is, the graph of the equation $y = \tan x$) turns out to be periodic with period π radians (or 180°). It consists of an infinite sequence of smooth curves crossing the x-axis at every multiple of π radians (180°) and having as asymptotes the vertical lines $x = \frac{k\pi}{2}$ radians ($x = k \times 90°$), for every odd integer k (which also means that the domain of the tangent function excludes these x-values).

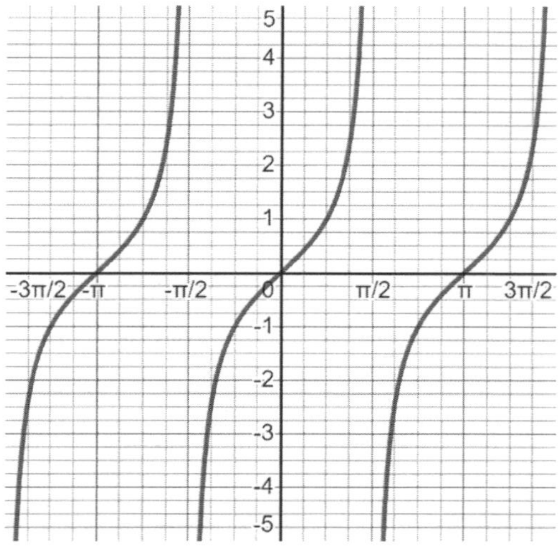

Review Video: Tangent
Visit mometrix.com/academy and enter code: 947639

Working with Functions

MANIPULATION OF FUNCTIONS

Translation occurs when values are added to or subtracted from the x- or y-values. If a constant is added to the y-portion of each point, the graph shifts up. If a constant is subtracted from the y-portion of each point, the graph shifts down. This is represented by the expression $f(x) \pm k$, where k is a constant. If a constant is added to the x-portion of each point, the graph shifts left. If a constant is subtracted from the x-portion of each point, the graph shifts right. This is represented by the expression $f(x \pm k)$, where k is a constant.

Stretching, compression, and reflection occur when different parts of a function are multiplied by different groups of constants. If the function as a whole is multiplied by a real number constant greater than 1, ($k \times f(x)$), the graph is stretched vertically. If k in the previous equation is greater than zero but less than 1, the graph is compressed vertically. If k is less than zero, the graph is reflected about the x-axis, in addition to being either stretched or compressed vertically if k is less than or greater than −1, respectively. If instead, just the x-term is multiplied by a constant greater than 1 ($f(k \times x)$), the graph is compressed horizontally. If k in the previous equation is greater than zero but less than 1, the graph is stretched horizontally. If k is less than zero, the graph is reflected about the y-axis, in addition to being either stretched or compressed horizontally if k is greater than or less than −1, respectively.

Review Video: Manipulation of Functions
Visit mometrix.com/academy and enter code: 669117

APPLYING THE BASIC OPERATIONS TO FUNCTIONS

For each of the basic operations, we will use these functions as examples: $f(x) = x^2$ and $g(x) = x$.

To find the sum of two functions f and g, assuming the domains are compatible, simply add the two functions together: $(f + g)(x) = f(x) + g(x) = x^2 + x$.

To find the difference of two functions f and g, assuming the domains are compatible, simply subtract the second function from the first: $(f - g)(x) = f(x) - g(x) = x^2 - x$.

To find the product of two functions f and g, assuming the domains are compatible, multiply the two functions together: $(f \times g)(x) = f(x) \times g(x) = x^2 \times x = x^3$.

To find the quotient of two functions f and g, assuming the domains are compatible, divide the first function by the second: $\left(\frac{f}{g}\right)(x) = \frac{f(x)}{g(x)} = \frac{x^2}{x} = x\,; x \neq 0$.

The example given in each case is fairly simple, but on a given problem, if you are looking only for the value of the sum, difference, product, or quotient of two functions at a particular x-value, it may be simpler to solve the functions individually and then perform the given operation using those values.

The composite of two functions f and g, written as $(f \circ g)(x)$ simply means that the output of the second function is used as the input of the first. This can also be written as $f(g(x))$. In general, this can be solved by substituting $g(x)$ for all instances of x in $f(x)$ and simplifying. Using the example functions $f(x) = x^2 - x + 2$ and $g(x) = x + 1$, we can find that $(f \circ g)(x)$ or $f(g(x))$ is equal to $f(x + 1) = (x + 1)^2 - (x + 1) + 2$, which simplifies to $x^2 + x + 2$.

It is important to note that $(f \circ g)(x)$ is not necessarily the same as $(g \circ f)(x)$. The process is not always commutative like addition or multiplication expressions. It *can* be commutative, but most often this is not the case.

EVALUATING LINEAR FUNCTIONS

A **function** can be expressed as an equation that relates an input to an output where each input corresponds to exactly one output. The input of a function is defined by the x-variable, and the output is defined by the y-variable. For example, consider the function $y = 2x + 6$. The value of y, the output, is determined by the value of the x, the input. If the value of x is 3, the value of y is $y = 2(3) + 6 = 6 + 6 = 12$. This means that when $x = 3$, $y = 12$. This can be expressed as the ordered pair (3,12).

It is common for function equations to use the form $f(x) =$ instead of $y =$. However, $f(x)$ and y represent the same thing. We read $f(x)$ as "f of x." The expression "f of x" implies that the value of f depends on the value of x. The function used in the example above could be expressed as $y = 2x + 6$ or $f(x) = 2x + 6$. Both functions represent the same line when graphed.

Functions that are expressed in the form $f(x) =$ are evaluated in the same way the equations are evaluated in the form $y =$. For example, when evaluating the function $f(x) = 3x - 2$ for $f(6)$, substitute 6 in for x, and simplify. In this case, $f(x) = 3x - 2$ becomes $f(6) = 3(6) - 2 = 18 - 2 = 16$. When x is 6, $f(x)$ is 16.

Example: To find the value of $f(8)$, calculate as follows:

$$f(x) = 3x - 2$$
$$f(8) = 3(8) - 2$$
$$f(8) = 22$$

Review Video: Evaluating Functions
Visit mometrix.com/academy and enter code: 588515

Advanced Functions

Step Functions

The double brackets indicate a step function. For a step function, the value inside the double brackets is rounded down to the nearest integer. The graph of the function $f_0(x) = [\![x]\!]$ appears on the left graph. In comparison $f(x) = 2\left[\!\!\left[\frac{1}{3}(x-1)\right]\!\!\right]$ is on the right graph. The coefficient of 2 shows that it's stretched vertically by a factor of 2 (so there's a vertical distance of 2 units between successive "steps"). The coefficient of $\frac{1}{3}$ in front of the x shows that it's stretched horizontally by a factor of 3 (so each "step" is three units long), and the $x - 1$ shows that it's displaced one unit to the right.

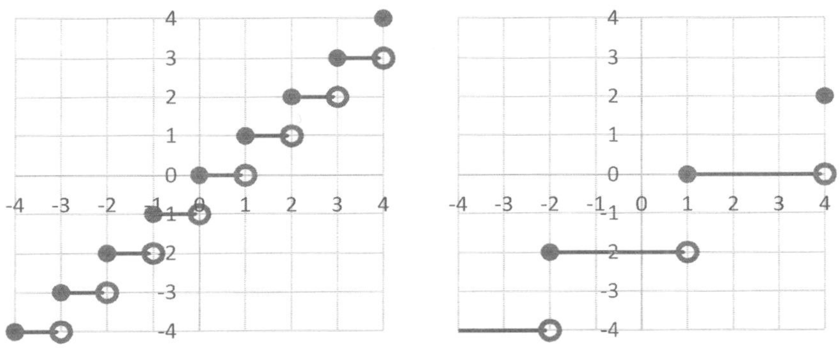

Transcendental Functions

Roughly speaking, algebraic functions are functions defined by formulas involving numbers and variables combined by addition, subtraction, multiplication, division, powers (but not variable powers), and roots. **Transcendental functions** are functions that are not algebraic. A function that includes logarithms, trigonometric functions, or variables as exponents, is transcendental, not algebraic, even if the function also includes polynomials or roots.

EXPONENTIAL FUNCTIONS

Exponential functions are functions that have the form $y = b^x$, where base $b > 0$ and $b \neq 1$. The exponential function can also be written $f(x) = b^x$. The following properties apply to exponential expressions:

Property	Description
$a^x a^y = a^{x+y}$	The product of exponentials with the same base equals the base raised to the sum of the powers
$a^x / a^y = a^{x-y}$	The quotient of exponentials with the same base equals the base raised to the difference of the powers
$(a^x)^y = a^{xy}$	An exponential raised to a power equals the base raised to the product of the powers
$(ab)^x = a^x b^x$	Exponentiation distributes over multiplication
$(a/b)^x = a^x / b^x$	Exponentiation distributes over division

The graph of an example exponential function, $f(x) = 2^x$, is below:

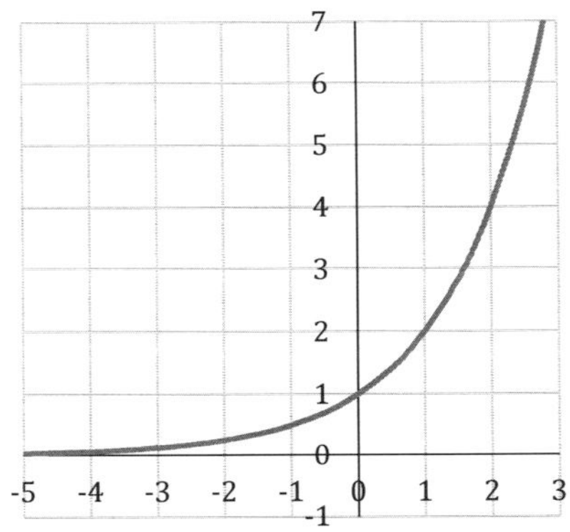

Note in the graph that the y-value approaches zero to the left and infinity to the right. One of the key features of an exponential function is that there will be one end that goes off to infinity and another that asymptotically approaches a lower bound. Common forms of exponential functions include the following:

Geometric sequences: The terms of a geometric sequence have the form $a_n = a_1 \times r^{n-1}$, where a_n is the value of the nth term, a_1 is the initial term, and r is the common ratio between succesive terms. Note that $a_1 \times r^{1-1} = a_1 \times r^0 = a_1 \times 1 = a_1$

Review Video: Geometric Sequences
Visit mometrix.com/academy and enter code: 140779

General exponential growth or decay: The general exponential growth or decay function is $f(t) = a(1 + r)^t$, where the value of $f(t)$ is the amount (of whatever quantity we are measuring) at time t, the constant a is the initial amount (the amount at time $t = 0$), and the constant r is the fixed rate of increase (if $r > 0$) or decrease (if $r < 0$) per unit time. For example, if we invest \$1000 at an interest rate of 6% interest compounded annually and we measure time in years, then our investment will grow according to an exponential model with $a = \$1000$ and $r = 6\% = 0.06$. Thus, after t years the value of our investment will be $f(t) = \$1000(1 + 0.06)^t$. After 5 years, for instance, the investment will grow to $f(5) = \$1000(1 + 0.06)^5 \approx \$1338.23$.

Compound interest: We can modify the general exponential growth function above slightly to get a formula for interest compounded n times per year. Replacing the constant a with P (for principal, the initial amount we invest), the total value of our investment after t years is $f(t) = P\left(1 + \frac{r}{n}\right)^{nt}$, where r is the nominal annual interest rate. Thus, if we change our $1000 investment above by having the interest rate be 6% compounded *semiannually* (two times per year—so $n = 2$), the value of our investment after t years is $f(t) = \$1000\left(1 + \frac{0.06}{2}\right)^{2t} = \$1000(1 + 0.03)^{2t}$. After 5 years, for instance, the investment will grow to $f(5) = \$1000(1 + 0.03)^{10} \approx \$1343.92$. This is slightly higher than the previous investment because the interest compounds twice as often.

Review Video: Compound Interest Formula
Visit mometrix.com/academy and enter code: 100366

Review Video: Interest Functions
Visit mometrix.com/academy and enter code: 559176

Population growth and continuously compounded interest: When a quantity grows (or decays) continuously at a rate that stays constant relative to the size the quantity has already attained, then we can model its size at time t with the function $f(t) = ae^{rt}$, where a is the initial amount, r is the relative growth rate (which we often call simply the **growth rate**), and e is the irrational constant known as Euler's number (approximately 2.718; scientific calculators have a button for finding e^x, which is known as the **natural exponential function**). For instance, under some circumstances, if an initial population (of people, plants, or animals, for example) of size a grows at constant relative rate r, then the size of the population at time t is $f(t) = ae^{rt}$. Similarly, if we invest principal P (instead of a) for t years at nominal annual interest rate r compounded *continuously*, then after t years the value of the investment is $f(t) = Pe^{rt}$.

For example, suppose the initial population of a town is $a = 1{,}200$ people and the relative annual growth rate is $r = 5\% = 0.05$. Then the population of the town after t years is $f(t) = ae^{rt} = 1200e^{0.05t}$. After 10 years, for instance, the town population is $f(10) = 1200e^{0.05\cdot 10} = 1200e^{0.5} \approx 1978$ people.

Review Video: Population Growth
Visit mometrix.com/academy and enter code: 109278

Logarithmic Functions

The **logarithmic function base b** is the function $y = \log_b x$ or $f(x) = \log_b x$, where the base b may be any positive number except one. The most common bases for logarithms are base 10 (the **common logarithm**) and base e (the **natural logarithm**). We usually write the common logarithm as $y = \log x$ (that is, $\log x$, with no base listed, means $\log_{10} x$) and the natural logarithm as $y = \ln x$ (that is, $\ln x$ means $\log_e x$).

Exponential functions and logarithmic functions with the same base are inverse functions. That is, if $f(x) = b^x$, then $f^{-1}(x) = \log_b x$. This means that the two equations $y = b^x$ (exponential form) and $x = \log_b y$ (logarithmic form) express the same relationship between the quantities x and y. We often solve problems involving logarithms by rewriting them in exponential form, and vice versa. Also, because of this inverse relationship, logarithms and exponentials cancel each other. That is, $\log_b b^x = x$ and $b^{\log_b x} = x$.

The following properties apply to logarithmic expressions:

Property	Description
$\log_b 1 = 0$	The log of 1 is equal to 0 for any base
$\log_b b = 1$	The log of the base is equal to 1
$\log_b b^p = p$	The log of the base raised to a power is equal to that power
$\log_b MN = \log_b M + \log_b N$	The log of a product is the sum of the log of each factor
$\log_b \frac{M}{N} = \log_b M - \log_b N$	The log of a quotient is equal to the log of the dividend minus the log of the divisor
$\log_b M^p = p \log_b M$	The log of a value raised to a power is equal to the power times the log of the value

Logarithms are helpful in solving equations in which the variable appears in an exponent. For instance, consider the example above in which we model the population of a town by the function $f(t) = 1200e^{0.05t}$. Suppose we want to know how long it will take the population of the town to double to 2400 people from its original value of 1200. We find this by solving the equation $1200e^{0.05t} = 2400$. First, we divide both sides of the equation by 1200 to get $e^{0.05t} = 2$. Then we take natural logarithms of both sides to get $\ln e^{0.05t} = \ln 2$, which simplifies (because natural logarithms and natural exponentials are inverse functions) to $0.05t = \ln 2$. Finally, we divide both sides by 0.05 to get $t = (\ln 2)/0.05$, which we evaluate with a calculator to get $t \approx 13.9$ years.

The graph of an example logarithmic function, $f(x) = \log_2(x + 2)$, is below:

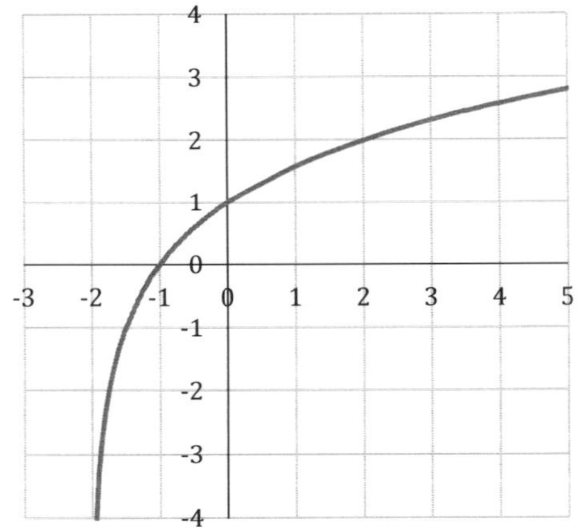

Review Video: Logarithmic Function
Visit mometrix.com/academy and enter code: 658985

Sequences

A **sequence** is an ordered set of numbers that continues in a defined pattern. The function that defines a sequence has a domain composed of the set of positive integers. Each member of the sequence is an element, or individual term. Each element is identified by the notation a_n, where a is the term of the sequence, and n is the integer identifying which term in the sequence a is.

There are two different ways to represent a sequence that contains the element a_n. The first is the simple notation $\{a_n\}$. The second is the expanded notation of a sequence: $a_1, a_2, a_3, \ldots a_n, \ldots$. Notice that the expanded form does not end with the nth term. There is no indication that the nth term is the last term in the sequence, only that the nth term is an element of the sequence.

Arithmetic Sequences

An **arithmetic sequence**, or arithmetic progression, is a special kind of sequence in which a specific quantity, called the common difference, is added to each term to make the next term. The common difference may be positive or negative. The general form of an arithmetic sequence containing n terms is $a_1, a_1 + d, a_1 + 2d, \dots, a_1 + (n-1)d$, where d is the common difference. The general formula for any term of an arithmetic sequence is $a_n = a_1 + (n-1)d$, where a_n is the term you are looking for and d is the common difference. To find the sum of the first n terms of an arithmetic sequence, use the formula $s_n = \frac{n}{2}(a_1 + a_n)$.

Review Video: Arithmetic Sequence
Visit mometrix.com/academy and enter code: 676885

Monotonic Sequences

A **monotonic sequence** is a sequence that is either nonincreasing or nondecreasing. A **nonincreasing** sequence is one whose terms either get progressively smaller in value or remain the same. Such a sequence is always bounded above, that is, all elements of the sequence must be less than some real number. A **nondecreasing** sequence is one whose terms either get progressively larger in value or remain the same. Such a sequence is always bounded below, that is, all elements of the sequence must be greater than some real number.

Recursive Sequences

When one element of a sequence is defined in terms of a previous element or elements of the sequence, the sequence is a **recursive sequence**. For example, given the recursive definition $a_1 = 1; a_2 = 1; a_n = a_{n-1} + a_{n-2}$ for all $n > 2$, you get the sequence 1,1,2,3,5,8, … . This is known as the Fibonacci sequence: a continuing sequence of numbers in which each number (after a_2) is the sum of the two previous numbers. The Fibonacci sequence can be defined as starting with either 1,1 or 0,1. Both definitions are considered correct in mathematics. Make sure you know which definition you are working with when dealing with Fibonacci numbers.

Sometimes in a recursive sequence, the terms can be found using a general formula that does not involve the previous terms of the sequence. Such a formula is called a **closed-form** expression for a recursive definition—an alternate formula that will generate the same sequence of numbers. However, not all sequences based on recursive definitions will have a closed-form expression. Some sequences will require the use of the recursive definition.

The Golden Ratio and the Fibonacci Sequence

The golden ratio is approximately 1.6180339887 and is often represented by the Greek letter phi, Φ. The exact value of Φ is $\frac{(1+\sqrt{5})}{2}$ and it is one of the solutions to $x - \frac{1}{x} = 1$. The golden ratio can be found using the Fibonacci sequence, since the ratio of a term to the previous term approaches Φ as the sequence approaches infinity:

n	a_n	a_{n-1}	$\frac{a_n}{a_{n-1}}$
3	2	1	2
4	3	2	1.5
5	5	3	$1.\overline{6}$
6	8	5	1.6
7	13	8	1.625
8	21	13	$1.\overline{615384}$
9	34	21	$1.\overline{619047}$
⋮	⋮	⋮	⋮
20	6,765	4,181	1.618033963 ...

Geometric Sequences

A geometric sequence is a sequence in which each term is multiplied by a constant number (called the common ratio) to get the next term. Essentially, it's the same concept as an arithmetic sequence, but with multiplication instead of addition.

Consider the following example of a geometric sequence: Andy opens a savings account with $10. During each subsequent week, he plans to double the amount from the previous week.

Sequence: $10, 20, 40, 80, 160, \ldots$

Function: $a_n = 10 \times 2^{n-1}$

This is a geometric sequence with a common ratio of 2. All geometric sequences represent exponential functions. The nth term in any geometric sequence is $a_n = a_1 \times r^{n-1}$, where a_n represents the value of the nth term, a_1 is the initial term, r is the common ratio, and n is the number of terms. Thus, substituting the initial value of 10 and common ratio of 2 gives the function $a_n = 10 \times 2^{n-1}$.

Review Video: Geometric Sequences
Visit mometrix.com/academy and enter code: 140779

The Unit Circle

Degrees, Radians, and the Unit Circle

It is important to understand the deep connection between trigonometry and circles. Specifically, the two main units, **degrees** (°) and **radians** (rad), that are used to measure angles are related this way: 360° in one full circle and 2π radians in one full circle: ($360° = 2\pi$ rad). The conversion factor relating the two is often stated as $\frac{180°}{\pi}$. For example, to convert $\frac{3\pi}{2}$ radians to degrees, multiply by the conversion factor: $\frac{3\pi}{2} \times \frac{180°}{\pi} = 270°$. As another example, to convert 60° to radians, divide by the conversion factor or multiply by the reciprocal: $60° \times \frac{\pi}{180°} = \frac{\pi}{3}$ radians.

Recall that the standard equation for a circle is $(x-h)^2 + (y-k)^2 = r^2$. A **unit circle** is a circle with a radius of 1 ($r = 1$) that has its center at the origin ($h = 0, k = 0$). Thus, the equation for the unit circle simplifies from the standard equation down to $x^2 + y^2 = 1$.

Standard position is the position of an angle of measure θ whose vertex is at the origin, the initial side crosses the unit circle at the point $(1, 0)$, and the terminal side crosses the unit circle at some other point (a, b). In the standard position, $\sin\theta = b$, $\cos\theta = a$, and $\tan\theta = \frac{b}{a}$.

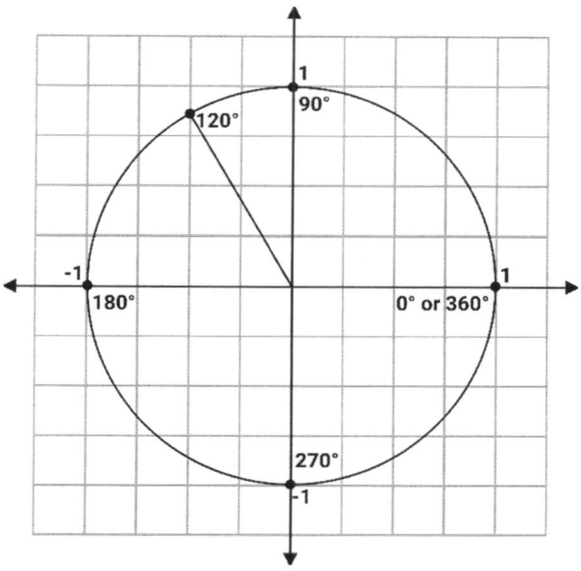

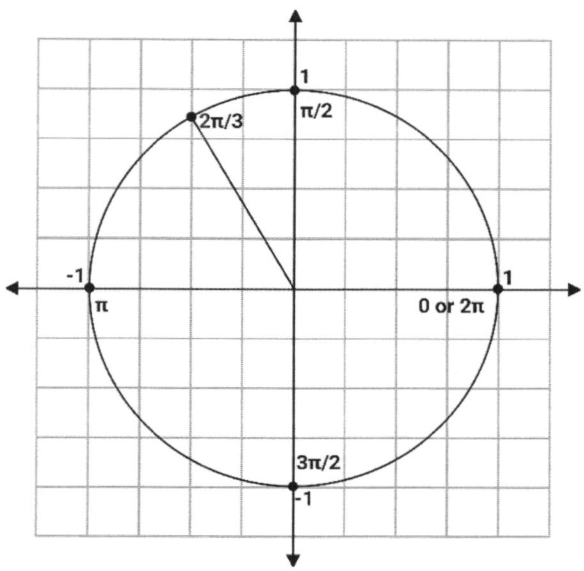

Review Video: Unit Circle
Visit mometrix.com/academy and enter code: 333922

TABLE OF COMMONLY ENCOUNTERED ANGLES

$0° = 0$ radians, $30° = \frac{\pi}{6}$ radians, $45° = \frac{\pi}{4}$ radians, $60° = \frac{\pi}{3}$ radians, and $90° = \frac{\pi}{2}$ radians

$\sin 0° = 0$	$\cos 0° = 1$	$\tan 0° = 0$
$\sin 30° = \frac{1}{2}$	$\cos 30° = \frac{\sqrt{3}}{2}$	$\tan 30° = \frac{\sqrt{3}}{3}$
$\sin 45° = \frac{\sqrt{2}}{2}$	$\cos 45° = \frac{\sqrt{2}}{2}$	$\tan 45° = 1$
$\sin 60° = \frac{\sqrt{3}}{2}$	$\cos 60° = \frac{1}{2}$	$\tan 60° = \sqrt{3}$
$\sin 90° = 1$	$\cos 90° = 0$	$\tan 90° = \text{undefined}$
$\csc 0° = \text{undefined}$	$\sec 0° = 1$	$\cot 0° = \text{undefined}$
$\csc 30° = 2$	$\sec 30° = \frac{2\sqrt{3}}{3}$	$\cot 30° = \sqrt{3}$
$\csc 45° = \sqrt{2}$	$\sec 45° = \sqrt{2}$	$\cot 45° = 1$
$\csc 60° = \frac{2\sqrt{3}}{3}$	$\sec 60° = 2$	$\cot 60° = \frac{\sqrt{3}}{3}$
$\csc 90° = 1$	$\sec 90° = \text{undefined}$	$\cot 90° = 0$

The values in the upper half of this table are values you should have memorized or be able to find quickly and those in the lower half can easily be determined as the reciprocal of the corresponding function.

Defined and Reciprocal Functions

The tangent function is defined as the ratio of the sine to the cosine: $\tan x = \frac{\sin x}{\cos x}$.

To take the reciprocal of a number means to place that number as the denominator of a fraction with a numerator of 1. The reciprocal functions are thus defined quite simply.

Cosecant	$\csc x$	$\frac{1}{\sin x}$
Secant	$\sec x$	$\frac{1}{\cos x}$
Cotangent	$\cot x$	$\frac{1}{\tan x}$

It is important to know these reciprocal functions, but they are not as commonly used as the three basic functions.

Review Video: Defined and Reciprocal Functions
Visit mometrix.com/academy and enter code: 996431

Inverse Trigonometric Functions

Inverse Functions

Each of the trigonometric functions accepts an angular measure, either degrees or radians, and gives a numerical value as the output. The inverse functions do the opposite; they accept a numerical value and give an angular measure as the output.

The inverse of sine, or arcsine, commonly written as either $\sin^{-1} x$ or $\arcsin x$, gives the angle whose sine is x. Similarly:

The inverse of $\cos x$ is written as $\cos^{-1} x$ or $\arccos x$ and means the angle whose cosine is x.
The inverse of $\tan x$ is written as $\tan^{-1} x$ or $\arctan x$ and means the angle whose tangent is x.
The inverse of $\csc x$ is written as $\csc^{-1} x$ or $\text{arccsc}\, x$ and means the angle whose cosecant is x.
The inverse of $\sec x$ is written as $\sec^{-1} x$ or $\text{arcsec}\, x$ and means the angle whose secant is x.

The inverse of $\cot x$ is written as $\cot^{-1} x$ or $\text{arccot}\, x$ and means the angle whose cotangent is x.

Review Video: Inverse Trig Functions
Visit mometrix.com/academy and enter code: 156054

Important Note About Solving Trigonometric Equations

When solving for an angle with a known trigonometric value, you must consider the sign and include all angles with that value. Your calculator will probably only give one value as an answer, typically in the following ranges:

- For $\sin^{-1} x$, $\left[-\frac{\pi}{2}, \frac{\pi}{2}\right]$ or $[-90°, 90°]$
- For $\cos^{-1} x$, $[0, \pi]$ or $[0°, 180°]$
- For $\tan^{-1} x$, $\left[-\frac{\pi}{2}, \frac{\pi}{2}\right]$ or $[-90°, 90°]$

It is important to determine if there is another angle in a different quadrant that also satisfies the problem. To do this, find the other quadrant(s) with the same sign for that trigonometric function and find the angle that has the same reference angle. Then check whether this angle is also a solution.

- In the first quadrant, all six trigonometric functions are positive.
- In the second quadrant, sin and csc are positive.
- In the third quadrant, tan and cot are positive.
- In the fourth quadrant, cos and sec are positive.

If you remember the phrase, "ALL Students Take Classes," you will be able to remember the sign of each trigonometric function in each quadrant. ALL represents all the signs in the first quadrant. The "S" in "Students" represents the sine function and its reciprocal in the second quadrant. The "T" in "Take" represents the tangent function and its reciprocal in the third quadrant. The "C" in "Classes" represents the cosine function and its reciprocal.

Trigonometric Identities

Sum and Difference

To find the sine, cosine, or tangent of the sum or difference of two angles, use one of the following formulas where α and β are two angles with known sine, cosine, or tangent values as needed:

$$\sin(\alpha \pm \beta) = \sin\alpha\cos\beta \pm \cos\alpha\sin\beta$$
$$\cos(\alpha \pm \beta) = \cos\alpha\cos\beta \mp \sin\alpha\sin\beta$$
$$\tan(\alpha \pm \beta) = \frac{\tan\alpha \pm \tan\beta}{1 \mp \tan\alpha\tan\beta}$$

HALF ANGLE

To find the sine or cosine of half of a known angle, use the following formulas where θ is an angle with a known exact cosine value:

$$\sin\left(\frac{\theta}{2}\right) = \pm\sqrt{\frac{(1 - cos\theta)}{2}}$$

$$\cos\left(\frac{\theta}{2}\right) = \pm\sqrt{\frac{(1 + \cos\theta)}{2}}$$

To determine the sign of the answer, you must recognize which quadrant the given angle is in and apply the correct sign for the trigonometric function you are using. If you need to find an expression for the exact sine or cosine of an angle that you do not know, such as sine 22.5°, you can rewrite the given angle as a half angle, such as $\sin\left(\frac{45°}{2}\right)$, and use the formula above:

$$\sin\left(\frac{45°}{2}\right) = \pm\sqrt{\frac{(1 - \cos(45°))}{2}} = \pm\sqrt{\frac{\left(1 - \frac{\sqrt{2}}{2}\right)}{2}} = \pm\sqrt{\frac{(2 - \sqrt{2})}{4}} = \pm\frac{1}{2}\sqrt{(2 - \sqrt{2})}$$

To find the tangent or cotangent of half of a known angle, use the following formulas where θ is an angle with known exact sine and cosine values:

$$\tan\frac{\theta}{2} = \frac{sin\theta}{1 + \cos\theta}$$

$$\cot\frac{\theta}{2} = \frac{\sin\theta}{1 - \cos\theta}$$

These formulas will work for finding the tangent or cotangent of half of any angle unless the cosine of θ happens to make the denominator of the identity equal to 0.

The Pythagorean theorem states that $a^2 + b^2 = c^2$ for all right triangles. The trigonometric identity that derives from this principle is stated in this way: $\sin^2\theta + \cos^2\theta = 1$.

Dividing each term by either $\sin^2\theta$ or $\cos^2\theta$ yields two other identities, respectively:

$$1 + \cot^2\theta = \csc^2\theta$$

$$\tan^2\theta + 1 = \sec^2\theta$$

Review Video: Sum and Difference Trigonometric Identities
Visit mometrix.com/academy and enter code: 468838

DOUBLE ANGLES

In each case, use one of the double angle formulas. To find the sine or cosine of twice a known angle, use one of the following formulas:

$$\sin(2\theta) = 2\sin\theta\cos\theta$$

$$\begin{aligned}\cos(2\theta) &= \cos^2\theta - \sin^2\theta \\ &= 2\cos^2\theta - 1 \\ &= 1 - 2\sin^2\theta\end{aligned}$$

To find the tangent or cotangent of twice a known angle, use the formulas where θ is an angle with known exact sine, cosine, tangent, and cotangent values:

$$\tan(2\theta) = \frac{2\tan\theta}{1-\tan^2\theta}$$

$$\cot(2\theta) = \frac{\cot\theta - \tan\theta}{2}$$

PRODUCTS

To find the product of the sines and cosines of two different angles, use one of the following formulas where α and β are two unique angles:

$$\sin\alpha\sin\beta = \frac{1}{2}[\cos(\alpha-\beta) - \cos(\alpha+\beta)]$$

$$\cos\alpha\cos\beta = \frac{1}{2}[\cos(\alpha+\beta) + \cos(\alpha-\beta)]$$

$$\sin\alpha\cos\beta = \frac{1}{2}[\sin(\alpha+\beta) + \sin(\alpha-\beta)]$$

$$\cos\alpha\sin\beta = \frac{1}{2}[\sin(\alpha+\beta) - \sin(\alpha-\beta)]$$

Review Video: Half-Angle, Double Angle, and Product Trig Identities
Visit mometrix.com/academy and enter code: 274252

COMPLEMENTARY

The trigonometric cofunction identities use the trigonometric relationships of complementary angles (angles whose sum is 90°). These are:

$$\cos x = \sin(90° - x)$$

$$\csc x = \sec(90° - x)$$

$$\cot x = \tan(90° - x)$$

Domain, Range, and Asymptotes in Trigonometry

The domain is the set of all possible real number values of x on the graph of a trigonometric function. Some graphs will impose limits on the values of x.

The range is the set of all possible real number values of y on the graph of a trigonometric function. Some graphs will impose limits on the values of y.

Asymptotes are lines that the graph of a trigonometric function approaches but never reaches. Asymptotes exist for values of x in the graphs of the tangent, cotangent, secant, and cosecant. The sine and cosine graphs do not have any asymptotes.

Domain, Range, and Asymptotes of the Six Trigonometric Functions

The domain, range, and asymptotes for each of the trigonometric functions are as follows:

- In the **sine** function, the domain is all real numbers, the range is $-1 \leq y \leq 1$, and there are no asymptotes.
- In the **cosine** function, the domain is all real numbers, the range is $-1 \leq y \leq 1$, and there are no asymptotes.
- In the **tangent** function, the domain is $x \in \mathbb{R}; x \neq \frac{\pi}{2} + k\pi$, the range is all real numbers, and the asymptotes are the lines $x = \frac{\pi}{2} + k\pi$.
- In the **cosecant** function, the domain is $x \in \mathbb{R}; x \neq k\pi$, the range is $(-\infty, -1]$ and $[1, \infty)$, and the asymptotes are the lines $x = k\pi$.
- In the **secant** function, the domain is $x \in \mathbb{R}; x \neq \frac{\pi}{2} + k\pi$, the range is $(-\infty, 1]$ and $[1, \infty)$, and the asymptotes are the lines $x = \frac{\pi}{2} + k\pi$.
- In the **cotangent** function, the domain is $x \in \mathbb{R}; x \neq k\pi$, the range is all real numbers, and the asymptotes are the lines $x = k\pi$.

In each of the above cases, k represents any integer.

Chapter Quiz

Ready to see how well you retained what you just read? Scan the QR code to go directly to the chapter quiz interface for this study guide. If you're using a computer, simply visit the online resources page at **mometrix.com/resources719/cchsalgebra** and click the Chapter Quizzes link.

Common Core Practice Test #1

1. $f(x) = 5x + 10$. If $x = 10$, then what is the value of $f(x)$?

a. 25
b. 60
c. 12
d. 5

2. The table below lists values for x and $f(x)$.

x	$f(x)$
1	2
2	5
3	10
4	17
5	26

Which of the following equations describes the relationship between x and $f(x)$?

a. $f(x) = x + 1$
b. $f(x) = x^2$
c. $f(x) = (-x)^2$
d. $f(x) = x^2 + 1$

3. What is $(x + 3)(2x - 5)$?

a. $2x^2 + x + 15$
b. $2x^2 + 11x - 15$
c. $2x^2 + x - 15$
d. $3x^2 + 11x + 15$

4. Mrs. Rose has 16 students in her class. Her class has three times as many girls as boys. How many girls and boys are in Mrs. Rose's class?

a. 12 girls, 4 boys
b. 4 girls, 12 boys
c. 3 girls, 1 boy
d. 9 girls, 7 boys

5. Which of the following is equivalent to $\sqrt{2^6}$?

a. 8
b. 32
c. 2
d. 4

6. Which of the following is equivalent to $\frac{24m^3n^2}{18m^2n^6}$?

a. $\frac{3m}{2n^4}$
b. $\frac{4mn^4}{3}$
c. $6m^5n^8$
d. $\frac{4m}{3n^4}$

7. Simplify: $3x(x + 6 - 2x + 1)$.

a. $-3x^2 + 21x$
b. $-3x^2 + 7$
c. $3x^2 - 6x^2 + 7$
d. $3x^2 - 21x$

8. Which of these relations represents a function?

a.

x	3	7	1	5
y	−3	−5	−2	−4

b.

x	2	4	2	1
y	−1	4	1	0

c.

x	−3	−3	−3	−3
y	1	2	3	4

d.

x	0	1	1	2
y	0	3	−3	6

9. If $f(x) = \frac{2}{3}x - 4$, what is the value of $f(-6)$?

a. −7
b. −4
c. −8
d. 0

10. What is the 6th term of this arithmetic sequence?

−6, −4, −2, 0, ...

a. 4
b. 2
c. −8
d. 6

11. If $f(x) = x^2 - x - 3$, what is the value of $f(-1)$?

a. −5
b. −1
c. −3
d. −2

12. Which of the following formulas for the nth term corresponds to this arithmetic sequence?

$-8, -4, 0, 4, 8, \ldots$

a. $a_n = 4n - 12$
b. $a_n = 4n - 4$
c. $a_n = -4n - 12$
d. $a_n = -4n - 4$

13. Which of the following figures contains a graph of the function $y = 2x + 2$?

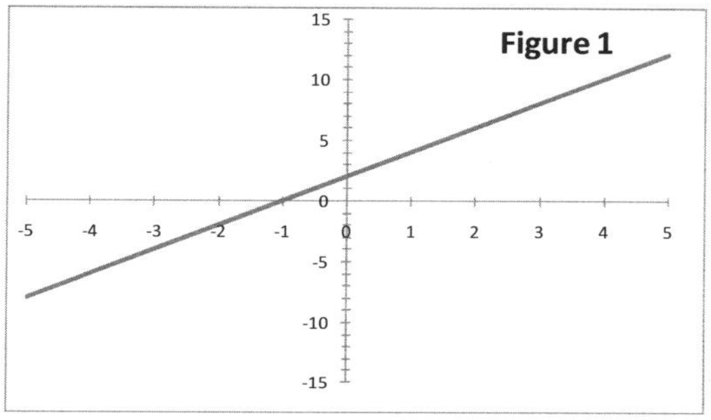

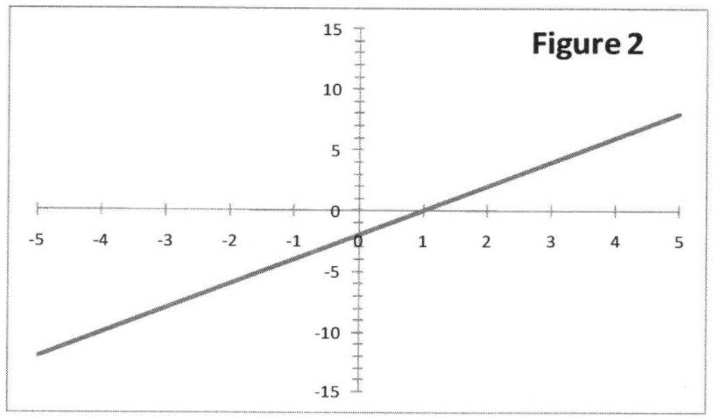

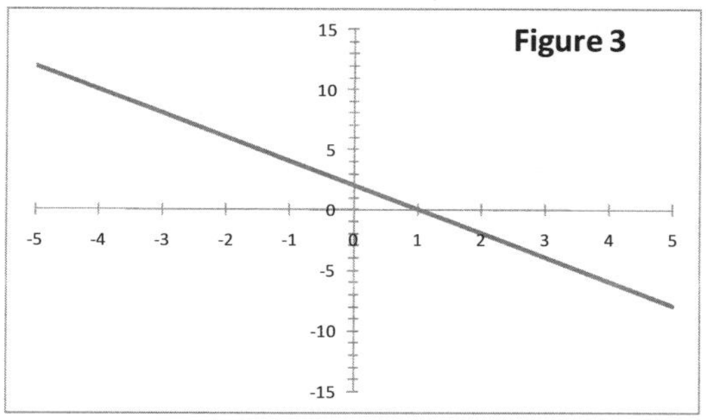

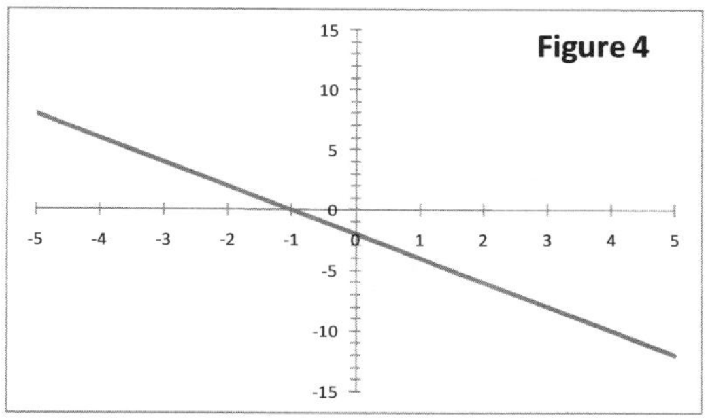

a. Figure 1
b. Figure 2
c. Figure 3
d. Figure 4

14. Which of the following figures contains a graph of the function $y = x^2 + 10$?

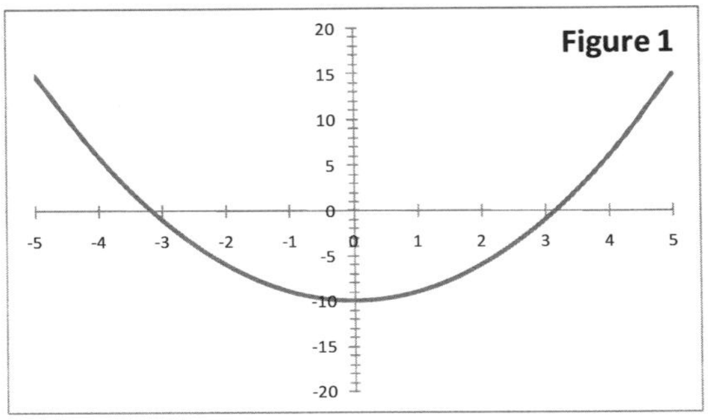

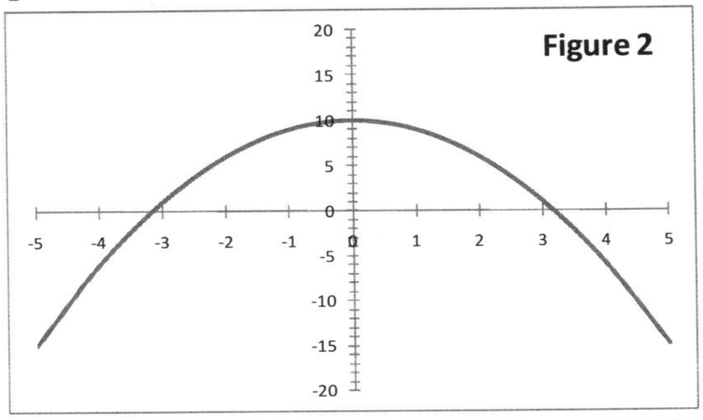

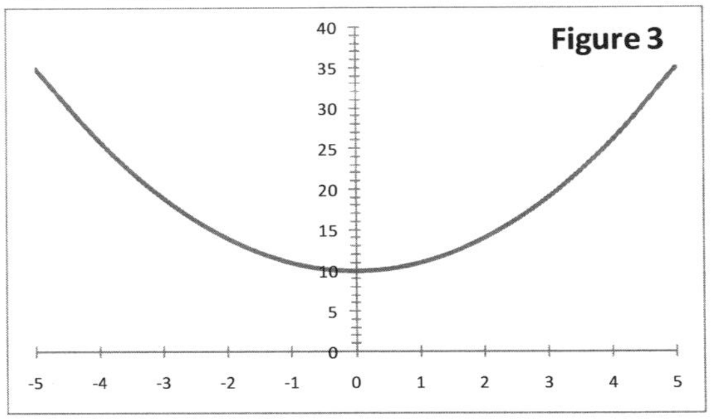

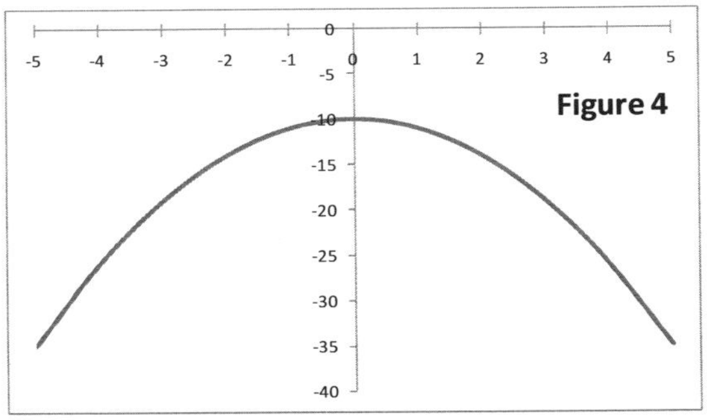

a. Figure 1
b. Figure 2
c. Figure 3
d. Figure 4

15. What is the correlation coefficient for the data in the table below?

The table shows the number of subs sold at Sam's Snack Shop the last several years.

Year	2010	2011	2012	2013	2014
Subs sold	222	263	302	345	394

a. 0.899
b. 0.999
c. 0.875
d. 0.937

16. One study involving 9th grade students compared the number of hours spent watching television with the number of hours slept each night. Analysis of the data revealed a correlation coefficient of −0.920. Which of the following conclusions can be drawn from these data?

a. Watching more television causes students to sleep more hours each night.
b. Watching more television causes students to sleep fewer hours each night.
c. Watching more television is associated with more hours slept each night.
d. Watching more television is associated with fewer hours slept each night.

17. This table shows the price of a gallon of milk over several years.

Years since 2000	10	11	12	13	14
Price of milk	$2.17	$2.36	$2.57	$2.79	$2.99

Which of the following equations represents the best-fit line for these data?

a. $y = 0.207x + 0.092$
b. $y = 0.312x + 0.142$
c. $y = 0.427x + 0.339$
d. $y = 0.524x + 0.465$

Questions 18 and 19 pertain to the following coordinate pairs:

$$\{(-5, 18), (-2, 12), (0, 3), (2, -3), (5, -12)\}$$

18. What is the domain of the coordinate pairs?

a. {18, 12, 3, –3, –12}
b. {-5, –2, 0, 2, 5}
c. {0, 3}
d. {-5, 18, 5, –12}

19. What is the range of the coordinate pairs?

a. {18, 12, 3, –3, –12}
b. {-5, –2, 0, 2, 5}
c. {0, 3}
d. {-5, 18, 5, –12}

Questions 20 and 21 pertain to the following scenario:

Aisha runs a small business selling candy bars to her classmates in school. She buys each candy bar for $0.75, and she sells each candy bar for $1.50. Let y represent Aisha's profit. Let x represent the number of candy bars she sells per day.

20. Which equation best represents Aisha's daily profit from selling candy bars?

a. $y = 0.75x - 1.50x$
b. $y = 0.75x + 0.75x$
c. $y = 1.50x + 0.75x$
d. $y = 1.50x - 0.75x$

21. Which figure contains the graph that best represents Aisha's daily profit from selling candy bars?

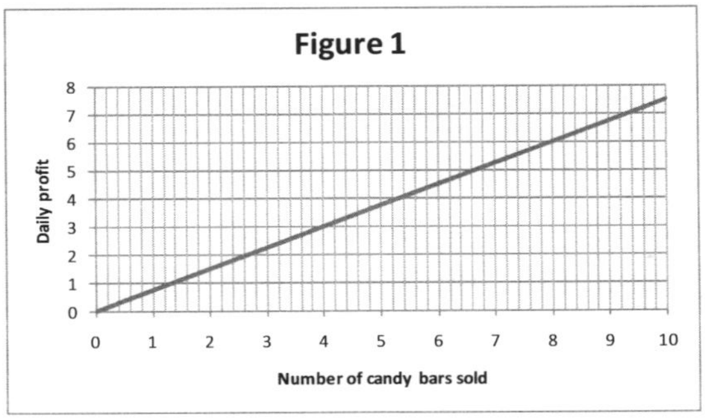

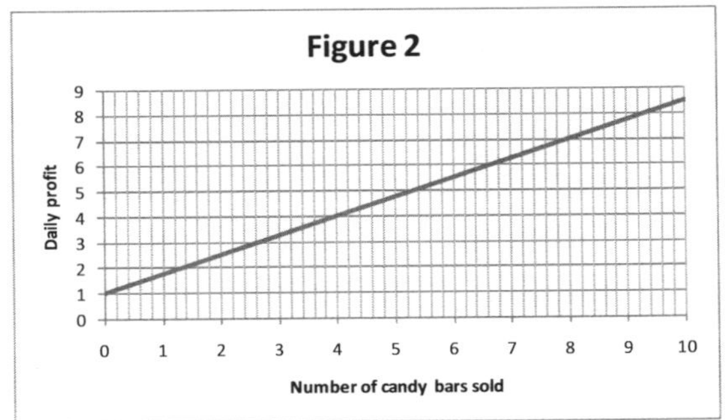

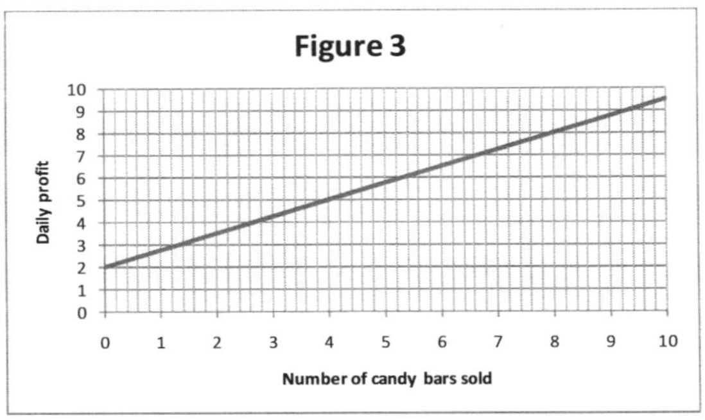

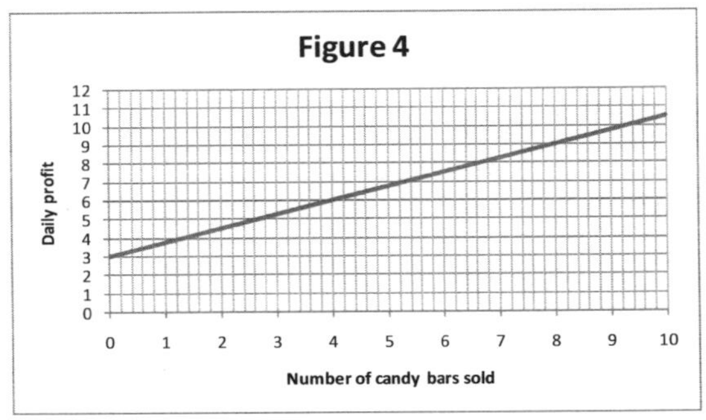

a. Figure 1
b. Figure 2
c. Figure 3
d. Figure 4

22. What are the factors of the following polynomial: $2x^2 + 7x - 15$?

a. $(2x + 5)(x - 3)$
b. $(x + 5)(2x - 3)$
c. $(2x - 5)(x + 3)$
d. $(x - 5)(2x + 3)$

23. What is the solution to the following equation: $x^2 - 9 = 0$?

a. $x = 3$
b. $x = -3$
c. Both A and B are solutions to the equation
d. Neither A nor B is a solution to the equation

24. What is the simplest form of the following polynomial?

$$4x^3 + x - x^3 + 2x^2 + 3 - 3x^3 + x - 2x^2 - 1$$

a. $2x + 2$
b. $x + 1$
c. $x^3 + 1$
d. $2(x + 1)$

25. Which of the following equations is an example of the distributive property?

a. $(5)(3) = (3)(5)$
b. $5 + 3 = 3 + 5$
c. $(5)(1 + 2) = (5)(1) + (5)(2)$
d. $15 = 15$

26. Which of the following binomials can be rewritten as a difference of squares with rational coefficients?

a. $16x^2 + 4$
b. $9x^2 - 12x + 4$
c. $4x^2 - y^2$
d. $8x^3 + 16$

27. Which of the following is the value of x when $3 < -\frac{1}{2}x + 1$?

a. $x < 4$
b. $x > 4$
c. $x < -4$
d. $x > -4$

28. John noticed that the number of points he scores during a basketball game is directly related to the number of hours he spends practicing each week. The table below lists John's weekly scores as a function of hours practiced. Let *h* represent the number of hours practiced and let p represent the number of points scored. Which equation represents the number of points John scored as a function of the number of hours he practiced?

Number of hours practiced	Number of points scored during basketball game
2	11
4	21
6	31
8	41
10	51

a. $p(h) = 5h + 1$
b. $p(h) = 5h - 1$
c. $p(h) = p + 10$
d. p(h)=p-10

29. Which of the following is parallel to the line $2x + 3y - 6 = 0$ and goes through the point $(6, 0)$?

a. $y = -\frac{2}{3}x + 4$
b. $y = -\frac{3}{2}x - 2$
c. $y = \frac{2}{3}x$
d. $y = \frac{3}{2}x - 7$

30. What is the equation of the line that is perpendicular to $y = 3x - 8$ and passes through the point (-3,2)?

a. $y = 3x + 11$
b. $y = \frac{1}{3}x + 3$
c. $y = -\frac{1}{3}x + 3$
d. $y = -\frac{1}{3}x + 1$

Questions 31 and 32 pertain to the following graph:

The graph describes the change in distance over time for a particular car.

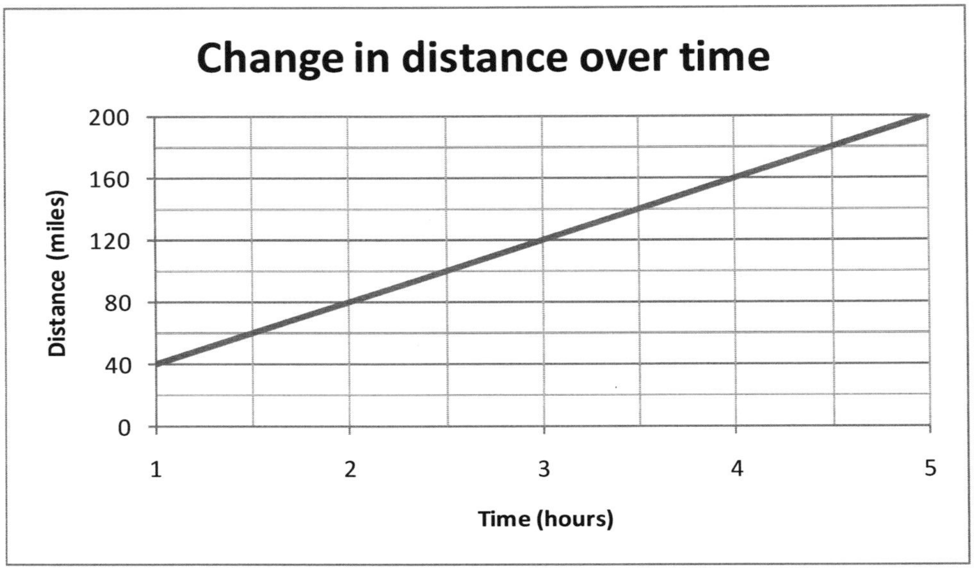

31. What is the slope of the line shown in the graph?

a. 20
b. 40
c. 60
d. 80

32. What are the units of the slope of the line?

a. time per distance
b. distance per time
c. hours per miles
d. miles per hour

Questions 33 and 34 pertain to the following information:

Let Equation A: $5y - 100x = 25$
Let Equation B: $5y - 200x = 75$

33. How does the slope of Equation B compare to the slope of Equation A?

a. The slope of Equation B is half the slope of Equation A.
b. The slope of Equation B is twice the slope of Equation A.
c. The slope of Equation B is 200 times the slope of Equation A.
d. The slope of Equation B is the same as the slope of the Equation A.

34. How does the y-intercept of Equation B compare to the y-intercept of Equation A?

a. The y-intercept of Equation B is twice the y-intercept of Equation A.
b. The y-intercept of Equation B is three times the y-intercept of Equation A.
c. The y-intercept of Equation B is 75 times the y-intercept of Equation A.
d. The y-intercept of Equation B is the same as the y-intercept of Equation A.

35. Line M contains the following two points: (1, 10) and (6, 20). What is the slope of line M?

a. 5
b. 2
c. 0.5
d. 10

36. Line Q has a slope of 10 and intercepts the y-axis at point (0, –15). What is the equation of line Q?

a. $y = -15$
b. $y = -15x + 10$
c. $y = 15x - 10$
d. $y = 10x - 15$

37. The equation for line 1 is $y_1 = 2x_1 + 6$ and the equation for line 2 is $y_2 = -x_2 - 3$. At what point does line 1 intersect line 2?

a. (–3, 6)
b. (6, –3)
c. (–3, 0)
d. (0,–3)

38. Table A below contains the *x*- and *y*-coordinates for several points on line *P*. Table B contains the *x*- and *y*-coordinates for several points on line *Q*. At what point does line *P* intersect line *Q*?

Table A: Coordinates for line *P*

x	y
−5	−8
−4	−4
−3	0
−2	4
−1	8
0	12
1	16

Table B: Coordinates for line *Q*

x	y
−5	4
−4	2
−3	0
−2	−2
−1	−4
0	−6
1	−8

a. (−3, 0)
b. (−5, −8)
c. (0, −6)
d. (1, −8)

39. If $f(x) = 3x - 8$, how would the function be affected by replacing $f(x)$ with $f(x + 2)$?

a. The graph would move two units to the right.
b. The graph would move two units to the left.
c. The graph would move two units to the right and two units up.
d. The graph would move two units to the left and two units up.

Questions 40 and 41 pertain to the following information:

Elli wants to plant a flower garden that contains only roses and tulips. However, she has a limited amount of space for the garden, and she can only afford to buy a specific number of each plant. Elli has enough space to plant a total of 20 flowers, and she has a total of $100 to purchase the flowers. Roses cost $14 per plant and tulips cost $4 per plant. Let *R* represent the number of roses and let *T* represent the number of tulips Elli will plant in her garden.

40. How many tulips will Elli plant in her flower garden?

a. 4
b. 20
c. 18
d. 2

41. Suppose the local greenhouse has a sale, allowing Elli to purchase roses for $9 per plant. Now how many roses and tulips will Elli plant in her garden?

a. 4 roses and 16 tulips
b. 16 roses and 4 tulips
c. 9 roses and 11 tulips
d. 20 roses and 0 tulips

42. Joshua has to earn more than 92 points on the state test in order to qualify for an academic scholarship. Each question is worth 4 points, and the test has a total of 30 questions. Let x represent the number of test questions. Which of the following inequalities can be solved to determine the number of questions Joshua must answer correctly?

a. $4x < 30$
b. $4x < 92$
c. $4x > 30$
d. $4x > 92$

43. A regular toilet uses 3.2 gallons of water per flush. A low flow toilet uses 1.6 gallons of water per flush. What is the difference between the number of gallons used by the regular toilet and the low flow toilet after 375 flushes?

a. 100 gallons
b. 525 gallons
c. 600 gallons
d. 1,200 gallons

Questions 44 and 45 pertain to the following information:

$$y_1 = x^2 \qquad y_2 = -x^2 \qquad y_3 = x^2 + 10$$

44. Which of the following numbers is included in the range of y_1?

a. 0
b. -1
c. -2
d. -3

45. How does function y_2 compare to the original function y_1?

a. y_2 has a different domain than y_1
b. y_2 has a different range than y_1
c. y_2 is shifted vertically by -1 unit when compared to y_1
d. y_2 is shifted horizontally by -1 unit when compared to y_1

46. A ball is dropped from a window 35 feet off the ground. Two seconds after dropping, the ball is 5 feet off the ground. If the height (h) in feet of the ball is modeled as a quadratic function of time (t) in seconds which of these functions represents this situation?

a. $h = -15t^2 + 30$
b. $h = -30t^2 + 35$
c. $h = -30t^2 + 30$
d. $h = -15t^2 + 35$

47. What is the equation of the parabola with vertex (0, 0) that passes through (−2, 8)?

a. $y = (2x)^2$
b. $y = 2(x + 2)^2$
c. $y = 2x^2$
d. $y = (x + 2)^2$

48. What is the equation of the parabola with vertex (–3, –6) that passes through (0, 3)?

a. $y = (x - 3)^2 - 6$
b. $y = (x - 6)^2 - 3$
c. $y = (x + 3)^2 - 6$
d. $y = (x + 3)^2 + 6$

49. What is the equation of a parabola that is symmetrical about the *y*-axis and goes through the points (0, –3) and (2, 5)?

a. $y = 2x^2 - 3$
b. $y = 2x^2$
c. $y = 3x^2 - 3$
d. $y = x^2 + 3$

50. At what value for *x* does the equation $x^2 + 10x = -25$ intercept the *x*-axis? Write your answer in the answer grid.

–	0	0
+	1	1
	2	2
	3	3
	4	4
	5	5
	6	6
	7	7
	8	8
	9	9

51. Consider the following equations:

$x^2 = 4$ $\quad x^3 = -8$ $\quad x^4 = 16$ $\quad x^5 = -32$

What is *x*? Write your answer in the answer grid.

–	0	0
+	1	1
	2	2
	3	3
	4	4
	5	5
	6	6
	7	7
	8	8
	9	9

52. Let y be inversely proportional to x such that $y = -\frac{1}{6}x$. If $y = 5$, what is x? Write your answer in the answer grid.

–	0	0
+	1	1
	2	2
	3	3
	4	4
	5	5
	6	6
	7	7
	8	8
	9	9

53. Solve the following equation for x:

$$2^x = 65{,}536$$

Write your answer in the answer grid.

–	0	0
+	1	1
	2	2
	3	3
	4	4
	5	5
	6	6
	7	7
	8	8
	9	9

54. What is a possible solution to $y = 3x^2 + 9x - 12$? Write your answer in the answer grid.

–	0	0
+	1	1
	2	2
	3	3
	4	4
	5	5
	6	6
	7	7
	8	8
	9	9

55. What are the domain and range of $y = \left(\frac{1}{2}\right)^x$?

a. The domain is $-2 \leq x$, and the range is $y < 2$.
b. The domain is $0 \leq x < +\infty$, and the range is $-2 < y < +\infty$.
c. The domain is $-\infty < x < +\infty$, and the range is $-\infty < y < +\infty$.
d. The domain is $-\infty < x < +\infty$, and the range is $0 < y < +\infty$.

56. Which of the following statements is true regarding the equation $y = 3(2)^x$?

a. This is a quadratic function.
b. This is a linear function.
c. This is an exponential growth function.
d. This is an exponential decay function.

57. Which of the following statements must be true about an exponential growth function in the form of $y = ab^x$?

a. $b > 1$
b. $b < 1$
c. $a > 1$
d. $a < 1$

58. What is the y-intercept of this graph of $y = 2(3^x)$?

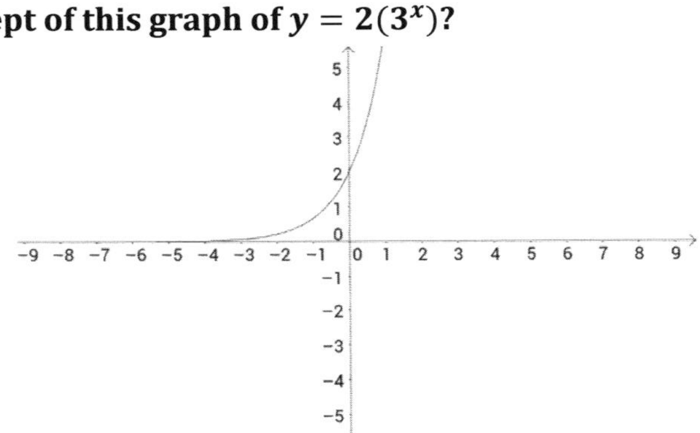

a. (0, 0)
b. (0, 2)
c. (2, 0)
d. (2, 2)

59. What is the y-intercept of the graph of $y = 6\left(\frac{1}{2}\right)^x$?

a. (0, −6)
b. (0, 3)
c. (6, 0)
d. (0, 6)

60. The population of a growing city in the 1940s is given in the table below. Use a graphing calculator to write an exponential function to model this growth. Which of the following is the best prediction of the population of this town in 1950?

Year	Population
1940	22,245
1941	23,357
1942	24,524
1943	25,750
1944	27,038

a. 34,710
b. 35,960
c. 36,230
d. 37,490

Answer Key and Explanations for Test #1

1. B: The equation describes a functional relationship between x and $f(x)$. To solve the equation, substitute 10 as the value of x, such that $f(10) = 5(10) + 10 = 50 + 10 = 60$. Therefore, choice B is correct.

2. D: For each value of x, $f(x) = x^2 + 1$,

$$f(1) = (1)^2 + 1 = (1)(1) + 1 = 1 + 1 = 2$$
$$f(2) = (2)^2 + 1 = (2)(2) + 1 = 4 + 1 = 5$$
$$f(3) = (3)^2 + 1 = (3)(3) + 1 = 9 + 1 = 10$$
$$f(4) = (4)^2 + 1 = (4)(4) + 1 = 16 + 1 = 17$$
$$f(5) = (5)^2 + 1 = (5)(5) + 1 = 25 + 1 = 26$$

Therefore, choice D is correct.

3. C: Applying the distributive property results in $x(2x) + x(-5) + 3(2x) + 3(-5)$. Simplifying the expression yields $2x^2 - 5x + 6x - 15$. Collecting like terms yields $2x^2 + x - 15$. Therefore, choice C is correct.

4. A: Let x represent the number of boys in Mrs. Rose's class. Since Mrs. Rose has three times as many girls in her class as boys, $3x$ represents the number of girls in Mrs. Rose's class. The total number of students in the class is 16. Written as an equation and solved for x we get:

$$x + 3x = 16$$
$$4x = 16$$
$$x = 4$$

Hence $x = 4$ and $3x = 12$. Therefore, 4 is the number of boys and 12 is the number of girls. Also,

$4 + 12 = 16$, the total number of students in the class. Therefore, choice A is correct.

5. A: The radical $\sqrt{2^6}$ can be rewritten as $2^{\frac{6}{2}}$ which is equal to 2^3 or 8. Therefore, choice A is correct.

6. D: When dividing expressions with exponents, the exponents of like bases are subtracted. The coefficients are divided. In this case, the fraction formed by the coefficients in the numerator and the denominator can be reduced. For the problem $\frac{24m^3n^2}{18m^2n^6}$, the quotient is $\frac{4m}{3n^4}$. Therefore, choice D is correct.

7. A: First simplify the terms inside the parentheses: $x + 6 - 2x + 1 = -x + 7$. Then use the distributive property to simplify, multiplying $3x$ by each term inside the parentheses: $3x(-x + 7) = -3x^2 + 21x$. Therefore, choice A is correct.

8. A: A relation is a function when each input has only one output. In this case, the relation is a function when each x-value has only one y-value. In choice B, $x = 2$ has two y-values. In choice C, $x = -3$ has four y-values. In choice D, $x = 1$ has two y-values. Therefore, choice A is correct.

9. C: To solve, substitute $x = -6$ into $f(x) = \frac{2}{3}x - 4$. Then $f(-6) = \frac{2}{3}(-6) - 4 = 2(-2) - 4 = -4 - 4 = -8$. Therefore, choice C is correct.

10. A: An arithmetic sequence increases or decreases at a constant rate. The intervals between the terms is a common difference. The common difference for this sequence is +2. The next three terms in this sequence are 2, 4, and 6. The 6th term is 4. Therefore, choice A is correct.

11. B: To solve, substitute $x = -1$ into $f(x) = x^2 - x - 3$. Then $f(-1) = (-1)^2 - (-1) - 3 = 1 + 1 - 3 = 2 - 3 = -1$. Therefore, choice B is correct.

12. A: The nth term of an arithmetic sequence is found by $a_n = a_1 + (n-1)d$ where d represents the common difference. The common difference for this sequence is +4. Then $a_n = -8 + (n-1)(4) = -8 + 4n - 4 = 4n - 12$. Therefore, choice A is correct.

13. A: The equation is written in the form of the point slope formula: $y = mx + b$ where m is the slope of the line and b is the y-axis intercept. For the given equation $y = 2x + 2$, the slope of the line is positive 2 and the line intercepts the y-axis at positive 2. The graph in Figure 1 fits these criteria. The graph in Figure 2 intercepts the y-axis at negative 2. The graphs in Figure 3 and Figure 4 have slopes of negative 2. Therefore, choice A is correct.

14. C: The equation is written in the form $y = Ax^2 + B$ where A tells the concavity of the graph and B is the y-intercept. In this case, A equals positive 1. So the graph is concave up. B equals positive 10. So the graph intercepts the y-axis at positive 10. The graph in Figure 3 fits these criteria. The graph in Figure 1 intercepts the y-axis at negative 10. The graphs in Figure 2 and Figure 4 are concave down. Therefore, choice C is correct.

15. B: Use a graphing calculator. Press STAT and select EDIT to enter the *year* data into List 1. Enter the *Subs sold* data into List 2. To perform the regression, press STAT and select CALC. Scroll down to LinReg(ax+b). The *r* value is the correlation coefficient. Therefore, choice B is correct. (Different graphing calculators might involve slightly different procedures; consult the instructions to your calculator.)

16. D: Since the correlation coefficient is close to −1, there is a strong negative correlation between the variables. Watching more television is associated with fewer hours slept each night. However, correlation does not imply causation. Therefore, choice D is correct.

17. A: Use a graphing calculator. Press STAT and select EDIT to enter the *Years since 2000* data into List 1. Enter the *Price of milk* data into List 2. To perform the regression, press STAT and select CALC. Scroll down to LinReg(ax+b). (Different graphing calculators might involve slightly different procedures; consult the instructions to your calculator.) With an equation in the form of $y = ax + b$, $a = 0.207$ and $b = 0.092$. Then the equation is $y = 0.207x + 0.092$. Therefore, choice A is correct.

18. B: The list of coordinate pairs represents the x- and y-values of five points. The domain is all the x-values. Answer B contains all the x-values of the coordinate pairs. Therefore, choice B is correct.

19. A: The list of coordinate pairs represents the x- and y-values of five points. The range is all the y-values. Answer A contains all the y-values of the coordinate pairs. Therefore, choice A is correct.

20. D: To calculate Aisha's daily profit, first determine the amount of money Aisha earns from selling candy. Since x represents the number of candy bars she sells per day, and she sells each bar for \$1.50, then her daily earnings equal $1.50x$. Next, determine how much money Aisha spends buying the candy. Since each bar costs \$0.75, she spends a total of $0.75x$ buying the candy. Finally, subtract the amount of money she spends buying the candy from the amount of money she earns selling the candy. Since y represents her daily profits, $y = 1.50x - 0.75x$. Therefore, choice D is correct.

21. A: The graph that represents Aisha's daily profit can be determined from the equation $y = 1.50x - 0.75x$. This equation can also be written as $y = (1.50 - 0.75)x$ or $y = 0.75x$. In its simplest form, the equation that describes Aisha's profits has a y-intercept of 0 and a slope of 0.75. The y-intercept tells the profits Aisha will earn if she sells no candy bars. Based on the equation, if Aisha sells no candy bars, then she earns no profit. The graph in Figure 1 fits these criteria. Therefore, choice A is correct.

22. B: To factor the polynomial, find factors of the first and third terms whose product can be added to get the middle term. The fastest way to find the correct answer is to multiply the answer choices and select the choice that yields the original equation. In this case,

$$\begin{aligned}(x + 5)(2x - 3) &= (x)(2x) + (x)(-3) + (5)(2x) + (5)(-3)\\ &= 2x^2 - 3x + 10x - 15\\ &= 2x^2 + 7x - 15\end{aligned}$$

Therefore, choice B is correct.

23. C: The solution to the equation follows:

$$\begin{aligned}x^2 - 9 &= 0\\ x^2 &= 9\\ x &= \sqrt{9}\end{aligned}$$

$$x = +3 \; and \; x = -3$$

Therefore, choice C is correct.

24. D: To simplify the polynomial, group and combine all terms of the same order.

$$\begin{aligned}&4x^3 + x - x^3 + 2x^2 + 3 - 3x^3 + x - 2x^2 - 1\\ &= (4x^3 - x^3 - 3x^3) + (2x^2 - 2x^2) + (x + x) + (3 - 1)\\ &= 0 + 0 + 2x + 2\\ &= 2(x + 1)\end{aligned}$$

Therefore, choice D is correct.

25. C: The distributive property says that terms inside a set of parentheses can be multiplied by a factor outside the parentheses. In other words, $a(b + c) = ab + ac$. Answer C fits this definition.

26. C: The difference of two squares is written in the form $a^2 - b^2$, which can be factored as $(a - b)(a + b)$. Of the answer choices, $4x^2 - y^2$ can be factored into $(2x - y)(2x + y)$. Therefore, choice C is correct.

27. C: To solve, we first subtract 1 from each side: $3 < -\frac{1}{2}x + 1 \rightarrow 2 < -\frac{1}{2}x$. Then multiply each side by –2 (note that changing the sign involves flipping the inequality): $2 < -\frac{1}{2}x \rightarrow -4 > x$. Then we can simply rewrite this as $x < -4$. To test, we can choose any number less than –4, such as –6, and substitute for x in the original equation: $3 < -\frac{1}{2}(-6) + 1 \rightarrow 3 < 3 + 1 \rightarrow 3 < 4$. This is true, so the equation seems to be true. We can further test by choosing a number outside the range, such as 0: $3 < -\frac{1}{2}(0) + 1$. Since 3 is not less than 1, this is not true, further verifying the answer. Therefore, choice C is correct.

28. A: In each case, the number of points scored p equals $5(h) + 1$ where h is the number of hours practiced. For example, $11 = (5)(2) + 1$ and $21 = (5)(4) + 1$. For answers C and D, the points scored are not written as functions of the hours practiced. Therefore, choice A is correct.

29. A: Two parallel lines have the same slope, so we first rearrange the original equation into point-slope form to find the slope. First, subtract $2x - 6$ from each side: $3y = -2x + 6$. Then divide each side by 3: $y = -\frac{2}{3}x + 2$. The slope is $-\frac{2}{3}$. Now we find the y-intercept by using the point (6, 0) in the slope-intercept equation $y = mx + b$: $0 = -\frac{2}{3}(6) + b$. Solving this yields $b = 4$, so the equation is $y = -\frac{2}{3}x + 4$. Therefore, choice A is correct.

30. D: When a line is perpendicular to another, its slope is the negative reciprocal of the other. The slope of the original equation is 3, so the slope of the perpendicular line is $-\frac{1}{3}$. To find the y-intercept, write the equation in slope-intercept form and substitute the given point: $2 = -\frac{1}{3}(-3) + b$, where b is the y-intercept. We solve for b: $2 = 1 + b$, so $b = 1$. The final equation is $y = -\frac{1}{3}x + 1$. Therefore, choice D is correct.

31. B: The slope of a line describes the change in the dependent variable divided by the change in the independent variable, i.e. the change in y over the change in x. To calculate the slope, consider any two points on the line. Let the first point be (1, 40) and let the second point be (2, 80):

$$\frac{y_2 - y_1}{x_2 - x_1} = \frac{80 - 40}{2 - 1} = \frac{40}{1} = 40$$

Therefore, choice B is correct.

32. D: The slope of the line is the change in y divided by the change in x. Therefore, the units of the slope are the units of y over the units for x. The unit for y is the unit for distance or miles. The unit for x is the unit for time or hours. Hence the units of the slope are miles over hours or miles per hour. Therefore, choice D is correct.

33. B: Write Equation B in slope-intercept form, which is $y = mx + b$:

$$\begin{aligned} 5y - 200x &= 75 \\ 5y &= 200x + 75 \\ y &= 40x + 15 \end{aligned}$$

Now write Equation A in slope-intercept form:

$$\begin{aligned} 5y - 100x &= 25 \\ 5y &= 100x + 25 \\ y &= 20x + 5 \end{aligned}$$

Since the slope Equation B is 40 and the slope of Equation A is 20, the slope of Equation B is twice the slope of Equation A. Therefore, choice B is correct.

34. B: Based on the slope-intercept form, which is $y = mx + b$, the y-intercept of Equation B is 15 and the y-intercept of Equation A is 5, so the y-intercept of Equation B is three time the y-intercept of Equation A. Therefore, choice B is correct.

35. B: The slope of a line is the change in y divided by the change in x. Calculate the slope as follows:

$$m = \frac{y_2 - y_1}{x_2 - x_1} = \frac{20 - 10}{6 - 1} = \frac{10}{5} = 2$$

Therefore, choice B is correct.

36. D: Write the equation in slope-intercept form: $y = mx + b$ where m is the slope of the line and b is the y-intercept. In this case, the slope $m = 10$ and the y-intercept $b = -15$. Hence $y = 10x - 15$. Therefore, choice D is correct.

37. C: At the intersection point of line 1 and line 2, $y_1 = y_2 = y$ and $x_1 = x_2 = x$. To find the x-coordinate, let $y_1 = y_2$.

$$\begin{aligned} 2x + 6 &= -x - 3 \\ 2x + x &= -6 - 3 \\ 3x &= -9 \\ x &= -3 \end{aligned}$$

Now find the y-coordinate by substituting $x = -3$ into either the equation for line 1 or the equation for line 2.

$$\begin{aligned} y &= 2x + 6 \\ y &= (2)(-3) + 6 \\ y &= -6 + 6 \\ y &= 0 \end{aligned}$$

Therefore, the point of intersection is (–3, 0) and choice C is correct.

38. A: The intersection point of line P and line Q will be common to both lines (see the explanation for question 34). Point (–3, 0) is the only point that is common to both lines. Therefore, choice A is correct.

39. B: The parent function is a line that intersects the y-axis at (0, 8). Substituting $x+2$ for x shifts this parabola two units to the left, but does not affect the y-intercept. Therefore, choice B is correct.

40. C: The total number of plants is 20, so we can write this as $R + T = 20$. Since roses cost \$14 each and tulips cost \$4 each, for a total cost of \$100, we can write this as $14R + 4T = 100$. We can solve this system of equations by substitution. First we need to rewrite the first equation:

$$R + T = 20$$

$$-R \quad = -R$$

$$T = 20 - R$$

Now we can combine the two equations, substituting the equation above for T:

$$14R + 4(20 - R) = 100$$

$$14R + 80 - 4R = 100$$
$$10R = 20$$
$$R = 2$$

Now we can solve for T by referring to our original first equation:

$$R + T = 20$$
$$2 + T = 20$$
$$T = 18$$

Thus, Elli will plant 18 tulips in her garden. Therefore, choice C is correct.

41. A: The new price for roses requires defining a new system of equations. Elli will still plant a total of 20 flowers. Hence

$$\begin{aligned} R + T &= 20 \\ T &= 20 - R \end{aligned}$$

However, based on the new price for roses,

$$\begin{aligned} 9R + 4T &= 100 \\ 9R + 4(20 - R) &= 100 \\ 9R + 80 - 4R &= 100 \\ 5R &= 20 \\ R &= 4 \end{aligned}$$

At the new price for roses, Elli will plant 4 roses in her garden. Since $R + T = 20$, Elli will plant 16 tulips in her garden. Therefore, choice A is correct.

42. D: In order to determine the number of questions Joshua must answer correctly, consider the number of points he must earn. Joshua will receive 4 points for each question he answers correctly, and x represents the number of questions. Therefore, Joshua will receive a total of $4x$ points for all the questions he answers correctly. Joshua must earn more than 92 points. Therefore, to determine the number of questions he must answer correctly, solve the inequality $4x > 92$.

43. C: To solve this problem, first calculate how many gallons each toilet uses in 375 flushes.

$$3.2 \times 375 = 1{,}200 \text{ gallons}$$
$$1.6 \times 375 = 600 \text{ gallons}$$

The problem is asking for the difference, so find the difference between the regular toilet and the low-flow toilet: $1{,}200 - 600 = 600$ gallons. Note that you could also find the difference in water use for one flush, and then multiply that amount by 375.

$$3.2 - 1.6 = 1.6$$
$$1.6 \times 375 = 600$$

With both methods of solving, the regular toilet uses 600 gallons more than the low flow toilet after 375 flushes.

44. A: The range is all the y-values. Refer to a graph of y_1 shown below.

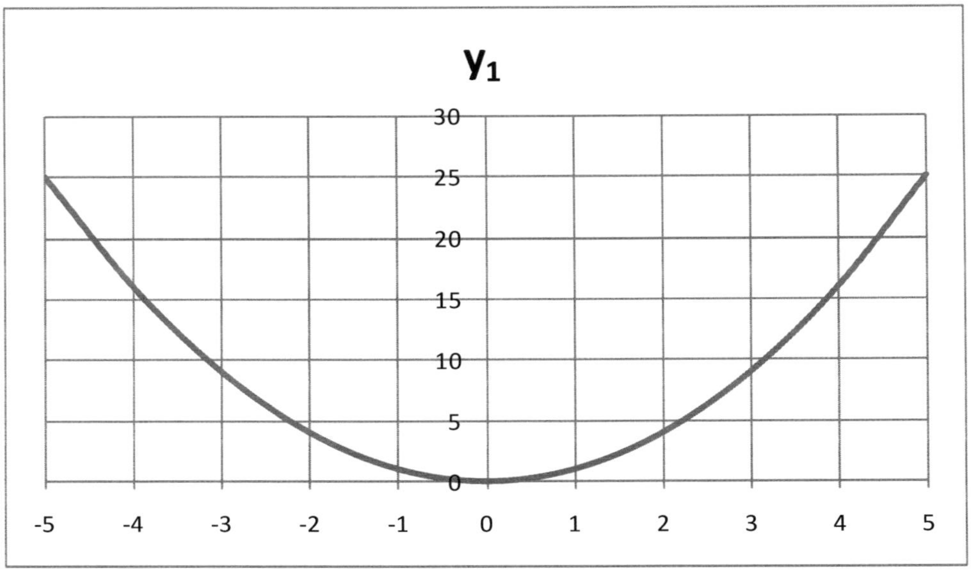

The minimum value for y is 0. Another way to solve this problem is to substitute the potential y-values in the equation for y_1. For example,

$$y_1 = x^2$$
$$-1 = x^2$$
$$\sqrt{-1} = x$$

This statement has no real solution since it requires taking the square root of a negative number. Similar solutions are obtained for $y = -2$ or $y = -3$. Therefore, choice A is correct.

45. B: The original function y_1 is concave up (see the graph of y_1 shown in the explanation for problem 46). Changing the coefficient of x^2 from +1 to −1 causes the function to be concave down. See the graph of y_2 shown below.

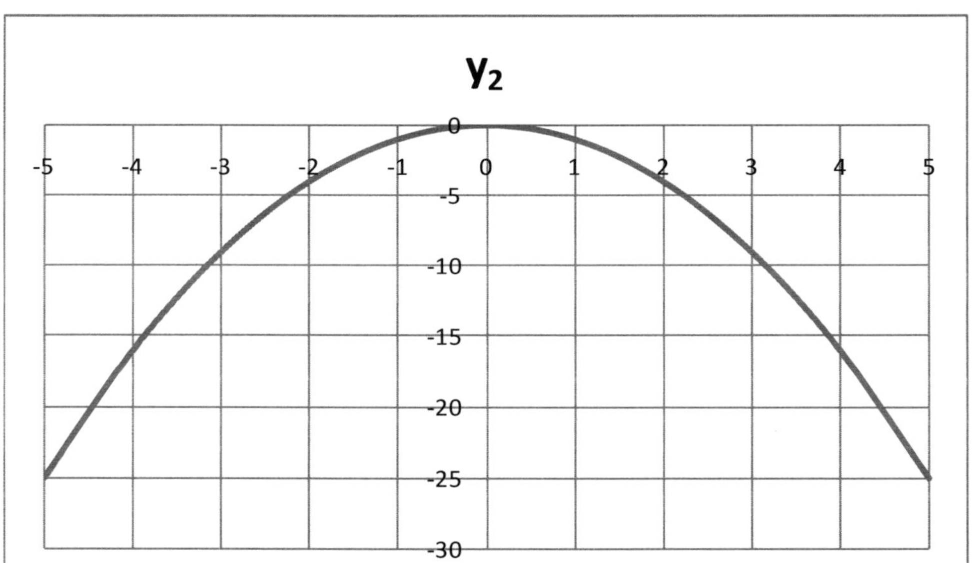

The y values for y_2 are different from those of y_1. Therefore, y_2 has a different range than y_1 and choice B is correct.

46. D: Since the ball drops from 35 feet, the vertex is at (0, 35). Since at 2 seconds after the fall, the height of the ball is 5 feet, the point (2, 5) also satisfies the function. The vertex form of a parabola is $y = a(x - h)^2 + k$ in which (h, k) is the vertex. Substituting in vertex (0, 35) for (h, k) and point (2, 5) for (x, y), $5 = a(2 - 0)^2 + 35$. Then $5 = 2a + 35$, and $a = -15$. The vertex form of this parabola is $y = -15x^2 + 35$ or $h = -15t^2 + 35$. Therefore, choice D is correct.

47. C: The vertex form of a parabola is $y = a(x - h)^2 + k$, in which (h, k) is the vertex. Substituting in the vertex (0, 0) for (h, k) and substituting the point (–2, 8) for (x, y), $8 = a(-2 - 0)^2 + 0$. Then $8 = 4a$, and $a = 2$. The vertex form of this parabola is $y = 2x^2$. Therefore, choice C is correct.

48. C: The vertex form of a parabola is $y = a(x - h)^2 + k$, in which (h, k) is the vertex. Substituting in vertex (–3, –6) for (h, k) and substituting the point (0, 3) for (x, y), $3 = a(0 + 3)^2 - 6$. Then $3 = 9a - 6$, and $a = 1$. The vertex form of this parabola is $y = (x + 3)^2 - 6$. Therefore, choice C is correct.

49. A: The vertex form of a parabola is $y = a(x - h)^2 + k$, in which (h, k) is the vertex. Because we are told that the parabola is symmetrical about the y-axis and that it goes through the point (0, –3), we know that (0, –3) is the vertex. Substituting this point for (h, k) and substituting the point (2, 5) for (x, y), $5 = a(2 - 0)^2 - 3$. Then $8 = 4a$, and $a = 2$. The vertex form of this parabola is $y = 2x^2 - 3$. Therefore, choice A is correct.

50. The correct answer is –5. The x-intercept is determined by setting the equation equal to zero and then solving for x. When $x^2 + 10x + 25 = 0$, then $x = -5$.

51. The correct answer is –2. Because $x^2 = 4$, we can quickly determine that the magnitude of x is 2. The sign, however, alternates between positive and negative. Therefore, x must be –2. Note that multiplying a negative number by itself an even number of times yields a positive number. Multiplying a negative number by itself an odd number of times yields a negative number.

52. The correct answer is –30. The solution is:

$$\begin{aligned} y &= -\frac{1}{6}x \\ 6y &= -x \\ -6y &= x \\ (-6)(5) &= x \\ -30 &= x \end{aligned}$$

53. The correct answer is 16. The solution follows:

$$\begin{aligned} 2^x &= 65{,}536 \\ xlog2 &= \log 65{,}536 \\ x &= \frac{\log 65{,}536}{\log 2} \\ x &= 16 \end{aligned}$$

54. The correct answer is 1 or –4. First factor out 3: $y = 3x^2 + 9x - 12 = 3(x^2 + 3x - 4)$. Then factor within the parentheses: $x^2 + 3x - 4 = (x + 4)(x - 1)$. Both $x + 4$ and $x - 1$ can be set equal to zero to solve. So the solutions are –4 and 1.

55. D: This is an exponential decay function in the form of $y = ab^x$ where $a > 0$ and $0 < b < 1$. Exponential decay functions have domains of all real values or $-\infty \leq x \leq +\infty$, and ranges of all positive real values or $0 < y < +\infty$. Therefore, choice D is correct.

56. C: The equation $y = 3(2)^x$ is an exponential growth function in the form of $y = ab^x$ where $a > 0$ and $b > 1$. An exponential decay function is in the form of $y = ab^x$ where $a > 0$ and $0 < b < 1$. Linear functions only have exponents of 1. Quadratic functions have at least one term of degree 2. Neither linear functions nor quadratic functions have a variable as an exponent. Therefore, choice C is correct.

57. A: For an exponential growth function in the form of $y = ab^x$, $b > 1$. If $b < 1$, the function will decrease instead of increase and correspond to exponential decay rather than exponential growth. It doesn't matter whether a is greater than or less than 1, as long as a is not zero. Therefore, choice A is correct.

58. B: The y-intercept is the point where the graph crosses the y-axis. When giving coordinates for a point, the x-coordinate is given first followed by the y-coordinate. This graph crosses the y-axis at (0, 2). Therefore, choice B is correct.

59. D: The y-intercept of a graph is the point where the graph crosses the x-axis or where $x = 0$. Substituting $x = 0$ into $y = 6\left(\frac{1}{2}\right)^x$ yields $y = 6\left(\frac{1}{2}\right)^0$ or $y = 6$. The y-intercept is (0, 6). Therefore, choice D is correct.

60. C: Use a graphing calculator. Press STAT and select EDIT to enter the year values into List 1. (Use years since 1940: 0, 1, 2, 3, 4.) Enter the population values into List 2. To perform the regression, press STAT and select CALC. Scroll down to ExpReg. Hit ENTER. (Different graphing calculators might involve slightly different procedures; consult the instructions to your calculator.) The particular coefficients may vary slightly, but you should get something like $y = 22{,}245e^{0.0488x}$ or $y = 22{,}245(1.050)^x$. Substituting $x = 10$ (for the year 1950) yields $y = 22{,}245e^{0.0488x} = 36{,}238$ or $y = 22{,}245(1.050)^{10} = 36{,}235$. Therefore, choice C is correct.

Common Core Practice Test #2

1. Let $p(y) = \frac{4y}{2} + 5$. If $y = 4$, then what is the value of $p(y)$?

a. 9
b. 7
c. 13
d. 37

2. What is the solution to $(3x^2 + 4x - 1) + (2x^2 - 7x - 3)$?

a. $5x^2 + 3x - 4$
b. $5x^2 - 3x - 4$
c. $5x^2 - 3x + 25$
d. $6x^2 - 11x + 2$

3. What is the solution to $\left(5x^2 - 2x - 2\right) - (3x^2 + 4x - 1)$?

a. $2x^2 - 6x - 1$
b. $2x^2 - 6x - 3$
c. $2x^2 + 2x - 3$
d. $8x^2 - 6x - 1$

4. Which of the following is equivalent to $3(2x^2 - x - 4)$?

a. $6x^2 - 3x + 7$
b. $6x^2 - 3x + 12$
c. $5x^2 - 3x - 7$
d. $6x^2 - 3x - 12$

5. Multiply: $(3x - 1)(2x + 5)$.

a. $6x^2 + 7x + 5$
b. $6x^2 + 13x + 5$
c. $6x^2 - 18x - 5$
d. $6x^2 + 13x - 5$

6. Which of the following binomials can be rewritten as a difference of squares with rational coefficients?

a. $4x^2 - 20$
b. $9x^2 - 25$
c. $6x^2 - 100$
d. $4x^2 + 25$

7. Which of the following is equivalent to $2\sqrt{4^4}$?

a. 4
b. 8
c. 32
d. 16

8. Which of the following is equivalent to $(2x^3y)(4x^5y^3)$?

a. $6x^8y^4$
b. $8x^8y^4$
c. $8x^{15}y^3$
d. $6x^{15}y^3$

9. What is the 7th term of this geometric sequence?

$-1, 2, -4, 8, \ldots$

a. 64
b. −16
c. −32
d. −64

10. Which of these graph descriptions fits a relation that is a function?

a. A vertical line
b. A parabola that opens to the right
c. A horizontal line
d. A circle

11. If $f(x) = 3x^2 - 2x + 3$, what is the value of $f(-2)$?

a. 11
b. 19
c. -5
d. 21

12. What are the factors of the following polynomial: $x^2 - x - 56$?

a. $(x - 7)(x + 8)$
b. $(x + 7)(x - 8)$
c. $(x - 7)(x - 8)$
d. $(x + 7)(x + 8)$

13. Which of the following correctly solves the formula $A = \frac{1}{2}(b_1 + b_2)h$ for b_2?

a. $b_2 = \frac{2A}{hb_1}$
b. $b_2 = \frac{2h}{A} - b_1$
c. $b_2 = \frac{2A}{h} - b_1$
d. $b_2 = \frac{2A}{h} + b_1$

14. Which of the following sets of points describes a function?

a. $(0,1)(-3,2)(2,-1)(-3,-3)$
b. $(0,1)(-3,2)(2,1)(-4,7)$
c. $(0,1)(-3,2)(0,3)(-4,7)$
d. $(0,1)(-3,2)(-3,1)(2,-1)$

15. What is the slope of a line containing points $(-1, 3)$ and $(4, -7)$?

a. $-\frac{4}{3}$
b. -2
c. $-\frac{4}{5}$
d. 2

16. What is the rate of change represented in this table?

x	y
5	3
10	4
15	5

a. 3
b. $\frac{1}{3}$
c. 5
d. $\frac{1}{5}$

17. What is the slope of this line?

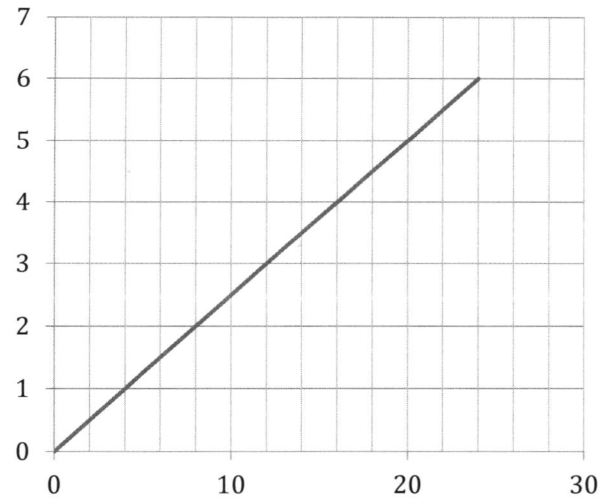

a. $-\frac{1}{4}$
b. 4
c. -4
d. $\frac{1}{4}$

18. Julia tracked growth of seedlings over a period of several days and noticed that in a week they grew 10.5 cm. What is the average rate of change?

a. 10.5 cm /day
b. 1.5 cm/day
c. 2.1 cm/day
d. 2.5 cm/day

19. Which of the following correctly graphs $3x - y < 4$?

a.
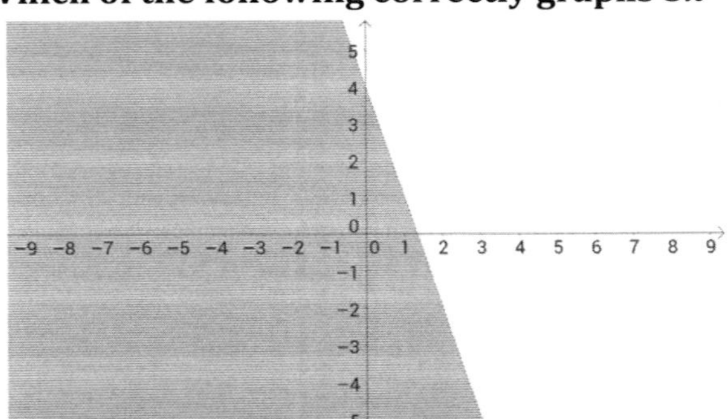

b.
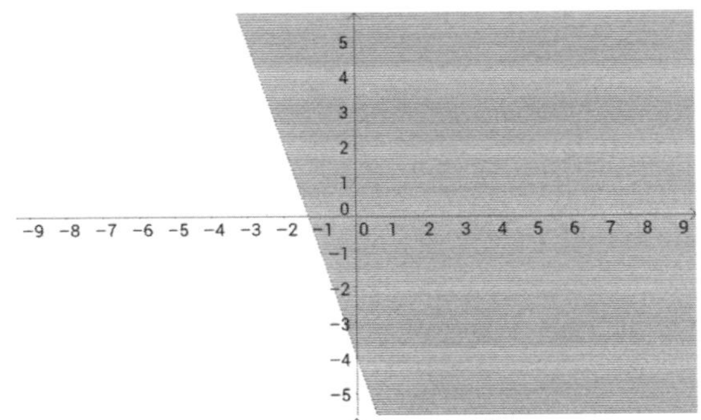

c.
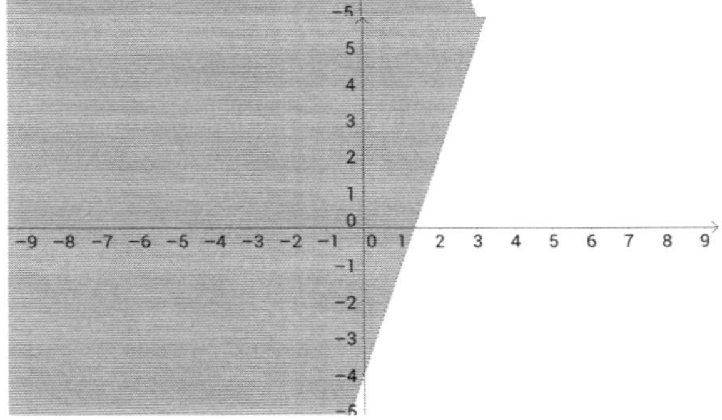

d.
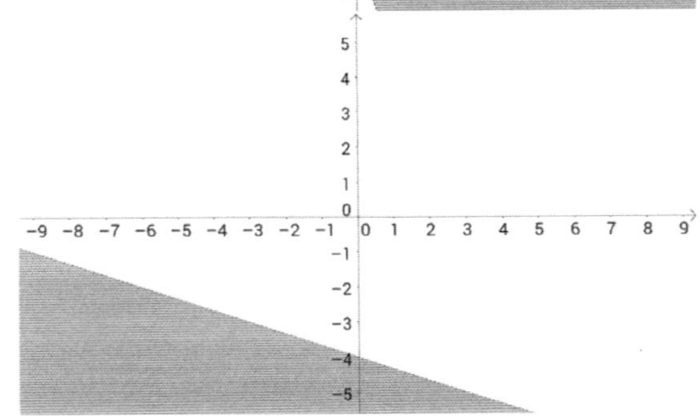

20. What is the effect on the graph of $f(x) = x$ when $f(x)$ is replaced by $2f(x)$?

a. The slope is doubled.
b. The slope is halved.
c. The graph is shifted up 2 units.
d. The graph is shifted down 2 units.

21. What is the solution to the system of equations represented by the graph below?

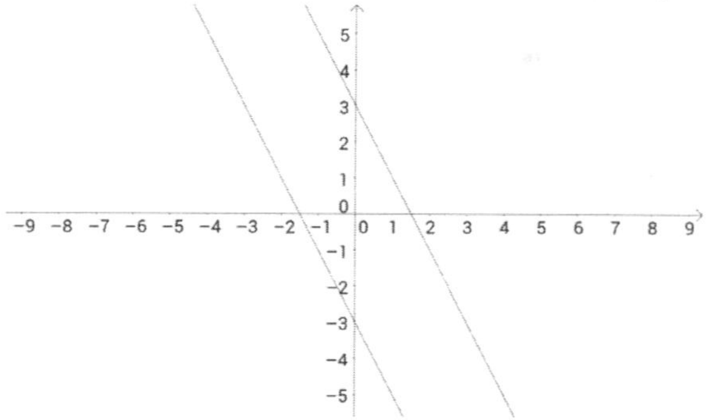

a. There are an infinite number of solutions.
b. There is no solution.
c. (0, 0)
d. (−3, 3)

22. Which of the following correctly graphs the system of linear inequalities of $y < 2x + 5$ and $y > x - 3$?

a.

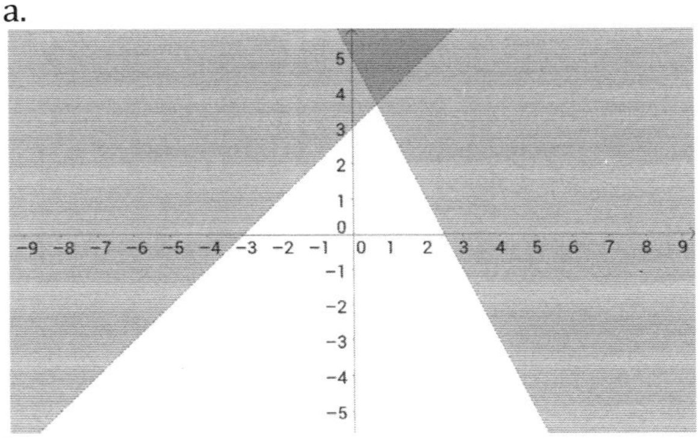

b.

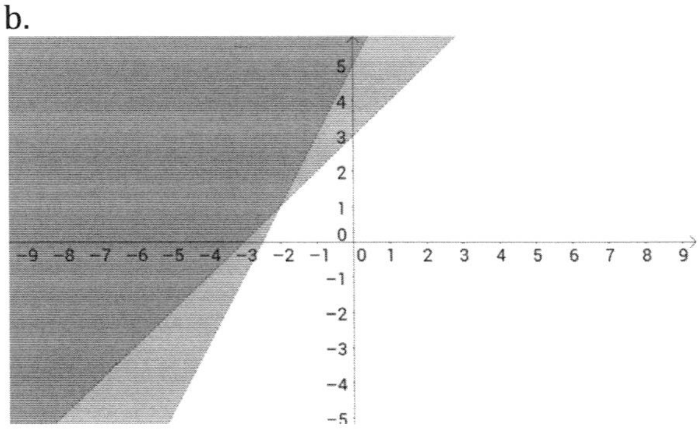

c.

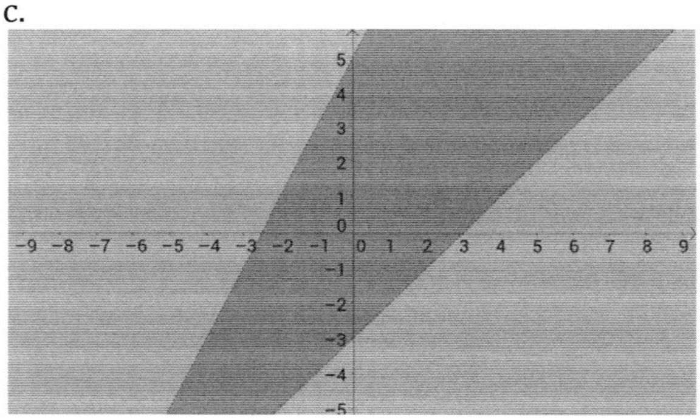

d.

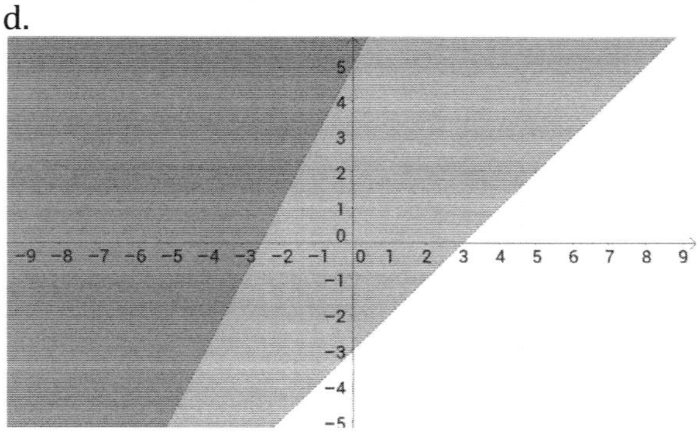

23. Several lab groups perform linear regressions for data sets recorded during a science experiment and calculate the corresponding correlation coefficients. Which lab group's equation for the best-fit line most closely models their data?

Data Set	Correlation coefficient
Steven's lab group	0.823
Sondra's lab group	0.500
Juan's lab group	0.132
Isabella's lab group	0.426

a. Steven's lab group
b. Sondra's lab group
c. Juan's lab group
d. Isabella's lab group

24. What is the equation of a line that is perpendicular to the y-axis and intersects the point (3,4)?

a. $y = 4$
b. $x = 3$
c. $y = \frac{4}{3}x$
d. $y = -\frac{3}{4}x + 4$

Question 25 pertains to the following chart:

Beth noticed that the number of questions she answers correctly on a test is directly related to the number of hours she spends studying. The table below lists her scores on five different tests as a function of hours spent studying. Let h represent the number of hours of study and let q represent the number of questions answered correctly.

Number of hours spent studying	Number of questions answered correctly
0.5	65
1	70
1.5	75
2	80
2.5	85

25. If the number of questions Beth answered correctly on a test were written as a linear function of the number of hours she studied, which set of numbers below would represent the domain of that function?

a. $\{65, 70, 75, 80, 85\}$
b. $\{0.5, 1, 1.5, 2, 2.5\}$
c. $\{0.5, 65\}$
d. $\{2.5, 85\}$

Questions 26 and 27 pertain to the following information:

Will is purchasing notebooks and boxes of pencils for his classroom. He wants to buy a total of 60 items, and he has a total of $100 to purchase the items. Notebooks cost $1.50 each and pencils cost $2 per box. Let N represent the number of notebooks and let P represent the number of boxes of pencils Will is going to purchase.

26. Which system of linear equations can be used to solve for the number of notebooks and pencils Will plans to purchase?

a. $\begin{cases} 1.5N + 2P = 60 \\ N + P = 100 \end{cases}$
b. $\begin{cases} N + P = 60 \\ 1.5N + 2P = 100 \end{cases}$
c. $\begin{cases} N + P = 60 \\ 2N + 1.5P = 100 \end{cases}$
d. $\begin{cases} 2N + 1.5P = 60 \\ 1.5N + 2P = 100 \end{cases}$

27. How many notebooks is Will planning to buy?

a. 15
b. 20
c. 40
d. 55

28. Which of the following is the value of x when $-3x + 1 < 2$?

a. $x < \frac{1}{3}$
b. $x > \frac{1}{3}$
c. $x < -\frac{1}{3}$
d. $x > -\frac{1}{3}$

29. The table shows the number of students enrolled in an afterschool taekwondo program by age groups.

Age	7	8	9	10
Enrollees in taekwondo	18	15	11	8

Which of the following equations represents the best-fit line for these data?

a. $y = -3.1x + 43.7$
b. $y = -3.4x + 41.9$
c. $y = -3.7x + 39.5$
d. $y = -3.9x + 37.3$

30. A study found a correlation coefficient of –0.143 between consumption of watermelon and heart disease. Which of the following conclusions can be drawn from these data?

a. Eating watermelon causes heart disease.
b. Eating watermelon helps prevent heart disease.
c. Eating watermelon is associated with a higher rate of heart disease.
d. Eating watermelon is not associated with a higher rate of heart disease.

31. The table shows the number of minutes a student practiced the violin the first four weeks of lessons. Based on these data, approximately how many minutes will the student practice week 7?

Week	1	2	3	4
Minutes practiced	20	37	49	55

a. 89 minutes
b. 91 minutes
c. 92 minutes
d. 93 minutes

32. What are the domain and range of the function represented by this table?

x	y
3	−3
−2	9
10	0
−5	4
7	−10

a. The domain is $\{-10, -5, -3, -2, 0\}$, and the range is $\{3, 4, 7, 9, 10\}$.
b. The domain is $\{3, 4, 7, 9, 10\}$, and the range is $\{-10, -5, -3, -2, 0\}$.
c. The domain is $\{-5, -2, 3, 7, 10\}$, and the range is $\{-10, -3, 0, 4, 9\}$.
d. The domain is $\{-10, -3, 0, 4, 9\}$, and the range is $\{-5, -2, 3, 7, 10\}$.

33. Which of the following is the equation for the line containing points (−3, −1) and (7, −2)?

a. $y = -\frac{3}{10}x + \frac{13}{10}$
b. $y = -\frac{1}{10}x - \frac{13}{10}$
c. $y = \frac{1}{10}x - \frac{7}{10}$
d. $y = -\frac{1}{10}x + \frac{13}{10}$

34. Which of the following is the linear equation representing the function given by this table?

x	y
−5	1
−4	3
−3	5
−2	7
−1	9

a. $y = -2x - 9$
b. $y = 2x + 11$
c. $y = 2x - 11$
d. $y = -2x + 9$

35. Which of the following equations is represented by this graph?

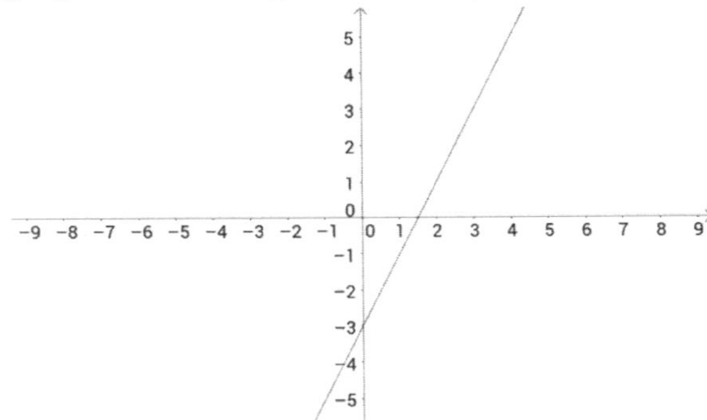

a. $y = 2x - 3$
b. $y = 3x - 3$
c. $y = x - 2$
d. $y = 2x - 5$

36. Given that y varies directly as x, if $y = 6$ when $x = 3$, which of these equations relates x to y?

a. $y = 2x$
b. $y = 6x$
c. $y = 3x$
d. $y = 9x$

37. Which of these equations represents a line that passes through point (−2, 1) and is parallel to $y = 3x - 1$?

a. $y = x - 2$
b. $y = -2x - 1$
c. $y = 3x + 7$
d. $y = 3x + 5$

38. Which of these equations represents a line that passes through point (3, −6) and that is perpendicular to $y = 4x + 3$?

a. $y = -4x - 6$
b. $y = -4x + 6$
c. $y = \frac{x}{4} - \frac{27}{4}$
d. $y = -\frac{x}{4} - \frac{21}{4}$

39. Which of the following is the equation for the line with a slope of −3 that passes through point (−1, 5)?

a. $y = 5x - 4$
b. $y = -x + 5$
c. $y = -3x + 2$
d. $y = -3x + 1$

40. What is the domain and the range of this function?

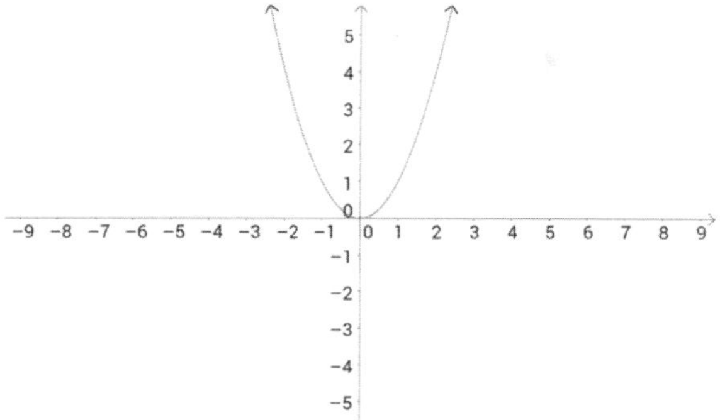

a. The domain is $x \leq 0$, and the range is $y \geq 0$.
b. The domain is $-\infty \leq x \leq +\infty$, and the range is $-\infty \leq y \leq +\infty$.
c. The domain is $x \geq 0$ and the range is $-\infty \leq y \leq +\infty$.
d. The domain is $-\infty \leq x \leq +\infty$, and the range is $y \geq 0$.

41. What are the domain and range of the equation graphed below?

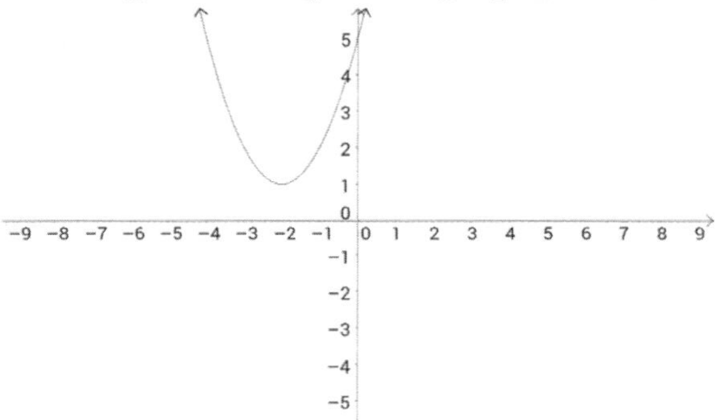

a. The domain is $-10 \leq x \leq +10$, and the range is $y \geq 1$.
b. The domain is $-\infty \leq x \leq +\infty$, and the range is $y \geq 1$.
c. The domain is $-10 \leq x \leq +10$, and the range is $y \geq 0$.
d. The domain is $-\infty \leq x \leq +\infty$, and the range is $y \geq 0$.

42. Match the following graphs to their respective functions: $y_1, y_2,$ and y_3.

$y_1 = x^2$
$y_2 = -x^2$
$y_3 = x^2 + 10$

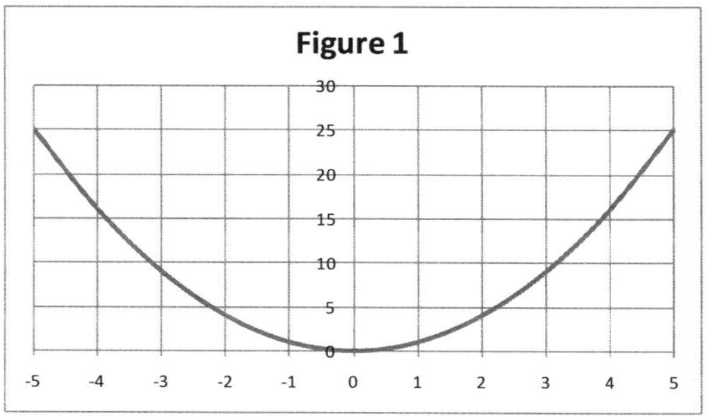

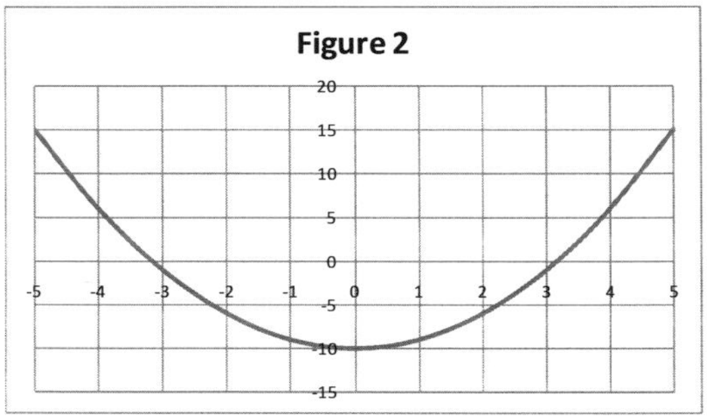

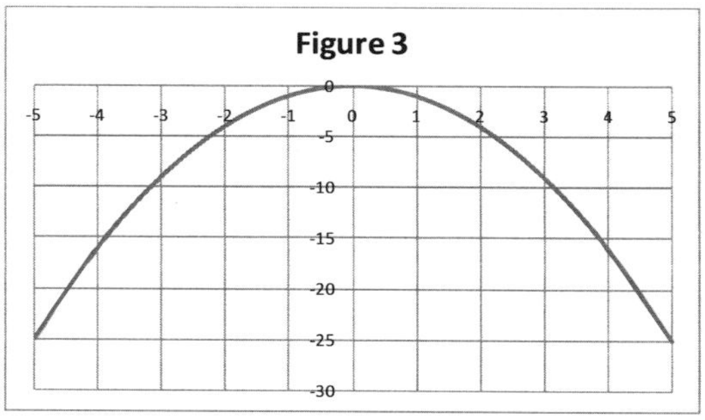

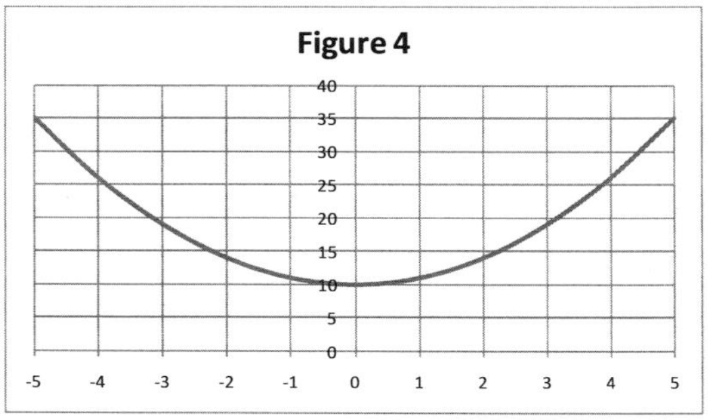

a. Figure 1 contains y_1. Figure 2 contains y_2. Figure 3 contains y_3.
b. Figure 2 contains y_1. Figure 3 contains y_2. Figure 4 contains y_3.
c. Figure 1 contains y_1. Figure 3 contains y_2. Figure 4 contains y_3.
d. Figure 3 contains y_1. Figure 2 contains y_2. Figure 1 contains y_3.

43. What is the y-intercept of the line described by Equation A: $5y - 100x = 25$? Write your answer in the answer grid.

-	0	0
+	1	1
	2	2
	3	3
	4	4
	5	5
	6	6
	7	7
	8	8
	9	9

44. Solve for A: $\frac{6+A}{5} = -2$ and write your answer in the answer grid.

-	0	0
+	1	1
	2	2
	3	3
	4	4
	5	5
	6	6
	7	7
	8	8
	9	9

45. Solve the following equation for x, and write your answer in the answer grid.

$$x^2 + 10x = -25$$

-	0	0
+	1	1
	2	2
	3	3
	4	4
	5	5
	6	6
	7	7
	8	8
	9	9

46. The axis of symmetry of this parabola is x = _______ (write your answer in the answer grid).

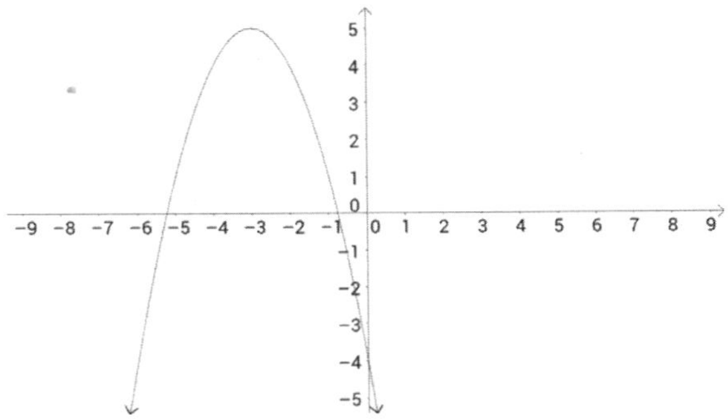

-	0	0
+	1	1
	2	2
	3	3
	4	4
	5	5
	6	6
	7	7
	8	8
	9	9

47. What is a possible solution to $3x^2 - 9x - 30 = 0$? Write your answer in the answer grid.

-	0	0
+	1	1
	2	2
	3	3
	4	4
	5	5
	6	6
	7	7
	8	8
	9	9

48. Which of the following equations is equivalent to $y = 3(x - 1)^2 + 4$?

a. $y = 3x^2 - 6x + 7$
b. $y = 3x^2 - 2x + 7$
c. $y = 3x^2 + 6x + 4$
d. $y = 2x^2 - 5x + 4$

49. Which of the following equations is equivalent to $y = -2(x + 5)^2 - 8$?

a. $y = 2x^2 + 20x + 42$
b. $y = -2x^2 + 20x + 42$
c. $y = -2x^2 - 20x - 50$
d. $y = -2x^2 - 20x - 58$

50. A rock falls from a 50-foot cliff. The height (h) in feet of the rock is modeled as a quadratic function of time (t) in seconds. If the rock is at 34 feet at 1 second after the fall, which of these functions represents this situation?

a. $h = -84t^2 + 50$
b. $h = -34t^2 + 50$
c. $h = 84t^2 + 50$
d. $h = -16t^2 + 50$

51. Which of the following statements is true regarding the linear factors of a quadratic equation?

a. The zeros of the quadratic equation can be determined directly from the linear factors.
b. The vertex of the quadratic equation can be determined directly from the linear factors.
c. The axis of symmetry of the quadratic equation can be determined directly from the linear factors.
d. The y-intercept of the quadratic equation can be determined directly from the linear factors.

52. What are the zeros of this quadratic function?

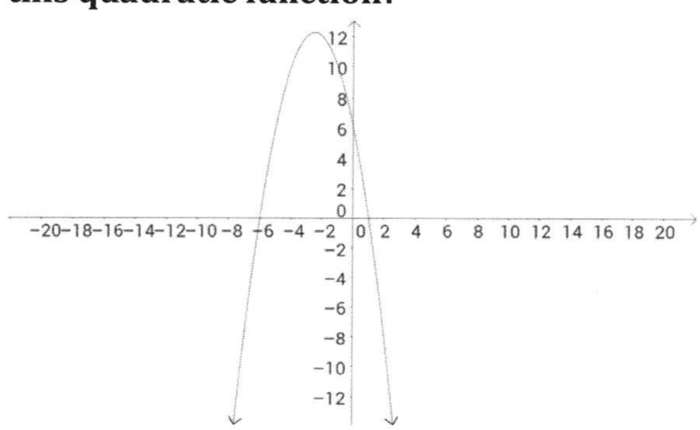

a. −2, 16
b. −6, 1
c. 0, 12
d. −2, 2

53. What is the solution to $2x^2 - x - 5 = 0$?

a. $x = \frac{1 \pm \sqrt{-39}}{4}$
b. $x = \frac{1 \pm \sqrt{-41}}{2}$
c. $x = \frac{1 \pm \sqrt{41}}{4}$
d. $x = \frac{-1 \pm \sqrt{39}}{4}$

54. What are the domain and range of $y = 2^x$?

a. The domain is $-2 \leq x$, and the range is $y < 2$.
b. The domain is $-\infty < x < +\infty$, and the range is $-2 < y < +\infty$.
c. The domain is $0 < x \leq +\infty$, and the range is $-\infty < y < +\infty$.
d. The domain is $-\infty < x < +\infty$, and the range is $0 < y < +\infty$.

55. What is the range of the exponential decay function pictured below?

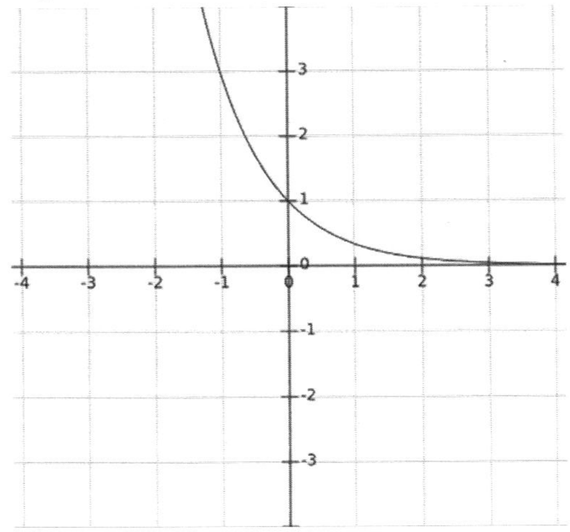

a. $-\infty < y < +\infty$
b. $0 < y < +\infty$
c. $-2 < y < +\infty$
d. $-\infty < x < +\infty$

56. Which of the following statements is true regarding the equation $y = 5\left(\frac{1}{4}\right)^x$?

a. This is an exponential decay function.
b. This is a linear function.
c. This is an exponential growth function.
d. This is a quadratic function.

57. Which of the following exponential functions models the growth of a town with a population of 20,000 with an annual increase of 2%?

a. $y = 20{,}001.02^x$
b. $y = 20{,}000(x)^{1.02}$
c. $y = 20{,}000(1.02)^x$
d. $y = 20{,}000(.02)^x$

58. Which of the following exponential functions models the growth of a $500 investment that is compounded annually with an interest rate of 4%?

a. $y = 500(4)^x$
b. $y = 500(x)^{1.04}$
c. $y = 5(400)^x$
d. $y = 500(1.04)^x$

59. Which of the following exponential functions models the decay in value of a $40,000 truck that loses 10% of its value every year?

a. $y = 500(4)^x$
b. $y = 40{,}000(0.90)^x$
c. $y = 5(400)^x$
d. $y = 500(1.04)^x$

60. Which of the following exponential functions provides a reasonable fit to this data?

x	y
0	57.2
5	28.5
10	14.2
15	7.08
20	3.52

a. $y = 57.22(0.870)^x$
b. $y = 56.99(0.560)^x$
c. $y = 3.52(0.615)^x$
d. $y = 28.5(5)^x$

Answer Key and Explanations for Test #2

1. C: The equation describes a functional relationship between y and $p(y)$. To solve the equation, substitute 4 as the value of y, such that

$$p(4) = \frac{4(4)}{2} + 5 = \frac{16}{2} + 5 = 8 + 5 = 13.$$

Therefore, choice C is correct.

2. B: When adding these polynomials, the parentheses can simply be removed, which results in $3x^2 + 4x - 1 + 2x^2 - 7x - 3$. Collecting like terms yields $5x^2 - 3x - 4$. Therefore, choice B is correct.

3. A: When subtracting polynomials, the parentheses around the first polynomial can simply be removed. When removing the parentheses around the second polynomial, the sign of each term must be switched. This results in $5x^2 - 2x - 2 - 3x^2 - 4x + 1$. Collecting like terms yields $2x^2 - 6x - 1$. Therefore, choice A is correct.

4. D: Applying the distributive property results in $3 \cdot 2x^2 - 3 \cdot x - 3 \cdot 4$. This expression can be simplified to $6x^2 - 3x - 12$. Therefore, choice D is correct.

5. D: Multiply each term in one set of parentheses by each term in the other set: $(3x)(2x) + (3x)(5) + (-1)(2x) + (-1)(5) = 6x^2 + 15x - 2x - 5$. Then combine like terms: $6x^2 + 13x - 5$. Therefore, choice D is correct.

6. B: A difference of squares factors to a sum and a difference. For example, $a^2 - b^2$ factors to $(a + b)(a - b)$. The binomial $9x^2 - 25$ can be rewritten as $(3x)^2 - (5)^2$. Then $(3x)^2 - (5)^2$ can be factored to $(3x + 5)(3x - 5)$. Therefore, choice B is correct.

7. C: The radical $2\sqrt{4^4}$ can be written as $2 \cdot 4^{\frac{4}{2}}$ which is equal to $2 \cdot 4^2$ or $2 \cdot 16$ or 32. Therefore, choice C is correct.

8. B: When multiplying expressions with exponents, the exponents of the like bases are added. The coefficients of the expressions are multiplied. For the problem $(2x^3y)(4x^5y^3)$, the product is $8x^8y^4$. Therefore, choice B is correct.

9. D: Consecutive terms in a geometric sequence have a common ratio. The common ratio of this sequence is –2. The next three terms are –16, 32, and –64. The 7th term is –64. Therefore, choice D is correct.

10. C: A function is a relation that has only one output for each input or only one y-value for each x-value. A vertical line drawn anywhere over the graph of a function will intersect the function at one point at most. Of the given choices, only a horizontal line fits this description. Each x-value of a horizontal line has only one y-value. Therefore, choice C is correct.

11. B: To solve, substitute $x = -2$ into $f(x) = 3x^2 - 2x + 3$. So $f(-2) = 3(-2)^2 - 2(-2) + 3 = 12 + 4 + 3 = 19$. Therefore, choice B is correct.

12. B: To factor the polynomial, find factors of the first and third term whose product can be added to get the middle term. Here, the factors 7 and –8 have a product of –56, and when added together yield −1. Another way to find the correct answer is to multiply the answer choices and select the choice that yields the original equation. In this case:

$$\begin{aligned}(x+7)(x-8) &= (x)(x)+(x)(-8)+(7)(x)+(7)(-8)\\ &= x^2-8x+7x-56\\ &= x^2-x-56\end{aligned}$$

Therefore, choice B is correct.

13. C: Multiplying both sides by 2 yields $2A=(b_1+b_2)h$. Then dividing both sides by h yields $\frac{2A}{h}=b_1+b_2$. Finally, subtracting b_1 from both sides yields $b_2=\frac{2A}{h}-b_1$. Therefore, choice C is correct.

14. B: A function has exactly one *y*-value for each *x*-value. Choices A, C, and D each have *x*-values with multiple *y*-values. Therefore, choice B is correct.

15. B: The slope can be found from $m=\frac{y_2-y_1}{x_2-x_1}$. Substituting the *x*-values and *y*-values from the given points yields $m=\frac{-7-3}{4-(-1)}=\frac{-10}{5}=-2$. The slope is −2. Therefore, choice B is correct.

16. D: The slope can be found from $m=\frac{y_2-y_1}{x_2-x_1}$. Substituting the *x*-values and *y*-values from the first two rows of the table yields $m=\frac{4-3}{10-5}=\frac{1}{5}$. The slope is $\frac{1}{5}$. Therefore, choice D is correct.

17. D: The slope is the change in *y* over the change in x, or the slope can be found from $m=\frac{y_2-y_1}{x_2-x_1}$. Since the line slants up to the right, the slope is positive. Selecting the points located at (12, 3) and (0, 0) and substituting the *x*-values and *y*-values into the formula yields $m=\frac{3-0}{12-0}=\frac{3}{12}$ or $\frac{1}{4}$. The slope is $\frac{1}{4}$. Therefore, choice D is correct. (Note that we didn't have to choose those two specific points; choosing any two points on the line would have resulted in the same answer.) Therefore, choice D is correct.

18. B: Rate of change, or slope, is calculated by dividing total change by total time. The seedlings grew 10.5 cm in 7 days, so the average rate of change is 10.5 ÷ 7, or 1.5 cm per day. Therefore, choice B is correct.

19. C: First, solving $3x-y<4$ for *y* results in $y>3x-4$. Then graph a dashed line with slope of 3 and *y*-intercept of −4. Select (0, 0) as a test point. Check if $0>3(0)-4$. Since this statement is true, shade the side of the line containing the point (0, 0). Therefore, choice C is correct.

20. B: The effect on the graph of *f*(*x*) = *x* when *f*(*x*) is replaced by 2*f*(*x*) is that the slope is halved. Solving the equation 2*f*(*x*) = *x* for *f*(*x*) yields $f(x)=\frac{x}{2}$. The slope of the line changes from 1 to $\frac{1}{2}$. Therefore, choice B is correct.

21. B: This system is of parallel lines. Since the lines never intersect, there is no solution. Therefore, choice B is correct.

22. C: First, graph a dashed line for $y=2x+5$. Using (0, 0) as a test point, check if $0=2(0)+5$. Since it's a true statement, shade the side of line $y=2x+5$ that includes (0, 0). Then, graph a dashed line for $y=x-3$. Using (0, 0) as a test point, check if $0>0-3$. Since it's a true statement, shade the side of the line $y=x-3$ that includes (0, 0). The area where the two shaded regions overlap is the solution to this system of inequalities. Therefore, choice C is correct.

23. A: The graphing calculator determines the best-fit line during the linear regression. The correlation coefficient measures how closely the equation of that best-fit line models the data. The closer the correlation coefficient is to 1 or −1, the better the model. Since Steven's lab group calculated a correlation coefficient of 0.823, which is the closest to 1, his lab group has the best model. Therefore, choice A is correct.

24. A: A line that is perpendicular to the *y*-axis is horizontal, with a slope of 0. Because it passes through the point (3, 4), we know that the *y*-value for the entire line is 4, and thus the equation is $y = 4$. Therefore, choice A is correct.

25. B: The domain consists of all the values of the independent variable. In this case, the independent variable is the number of hours spent studying. Therefore, choice B is correct.

26. B: Since Will is buying a total of 60 items, the number of notebooks plus the number of boxes of pencils is 60 or $N + P = 60$. Each notebook costs $1.50, so multiply the number of notebooks by 1.5. Each box of pencils costs $2, so multiply the number of boxes by 2. Will has a total of $100 to spend, so $1.5N + 2P = 100$. Therefore, choice B is correct.

27. C: Use a linear system of equations to find the number of notebooks. See the explanation for question 24. In this case, the system of equations is $N + P = 60$ and $1.5N + 2P = 100$. Begin with $N + P = 60$ and solve for *P*.

$$N + P = 60$$
$$P = 60 - N$$

Now substitute the equation for *P* into the equation $1.5N + 2P = 100$:

$$\begin{aligned} 1.5N + 2P &= 100 \\ 1.5N + 2(60 - N) &= 100 \\ 1.5N + 120 - 2N &= 100 \\ -0.5N &= -20 \\ N &= 40 \end{aligned}$$

Will is going to buy 40 notebooks. Therefore, choice C is correct.

28. D: To solve, we first subtract 1 from each side: $-3x < 2 - 1 \rightarrow -3x < 1$. Then divide each side by -3 (note that changing the sign involves flipping the inequality): $x > -\frac{1}{3}$. To test, we can choose any number greater than $-\frac{1}{3}$, such as 0, and substitute for *x* in the original equation: $-3(0) + 1 < 2 \rightarrow 1 < 2$. This is true, so the equation seems to be true. We can further test by choosing a number outside the range, such as -1: $-3(-1) + 1 < 2 \rightarrow 4 < 2$. Since 4 is not less than 2, this is not true, further verifying the answer. Therefore, choice D is correct.

29. B: Use a graphing calculator. Press STAT and select EDIT to enter the *Age* data into List 1. Enter the *Enrollees in taekwondo* data into List 2. To perform the regression, press STAT and select CALC. Scroll down to LinReg(*ax*+*b*). (Different graphing calculators might involve slightly different procedures; consult the instructions to your calculator.) With an equation in the form of $y = ax + b$, $a = -3.4$ and $b = 41.9$. Then the equation is $y = -3.4x + 41.9$. Therefore, choice B is correct.

30. D: Since the correlation coefficient is close to 0, there is a very weak negative correlation between the variables. Because correlation does not imply causation, the first two choices cannot be correct. Eating watermelon does not appear to have an association with rates of heart disease, due to the weak correlation. Therefore, choice D is correct.

31. D: Use a graphing calculator. Press STAT and select EDIT to enter the *Week* data into List 1. Enter the *Minutes practiced* data into List 2. To perform the regression, press STAT and select CALC. Scroll down to LinReg(*ax*+*b*). (Different graphing calculators might involve slightly different procedures; consult the instructions to your calculator.) With an equation in the form of $y = ax + b$, $a = 11.7$ and $b = 11$. The equation is $y = 11.7x + 11$. If $x = 7$, $y = (11.7)(7) + 11$ or approximately 93 minutes. Therefore, choice D is correct.

32. C: The domain is the set containing all of the inputs or *x*-values for the function. The range is the set containing all of the outputs or *y*-values for the function. The domain is {−5, −2, 3, 7, 10}, and the range is {−10, −3, 0, 4, 9}. Therefore, choice C is correct.

33. B: The slope can be found from $m = \frac{y_2 - y_1}{x_2 - x_1}$. Then $m = \frac{-2-(-1)}{7-(-3)} = \frac{-2+1}{7+3} = -\frac{1}{10}$. Substituting point (−3, −1) and this slope into $y = mx + b$ yields $-1 = \left(-\frac{1}{10}\right)(-3) + b$. Solving for b, $b = -1 - \frac{3}{10} = -\frac{13}{10}$. The equation of this line is $y = -\frac{1}{10}x - \frac{13}{10}$. Therefore, choice B is correct.

34. B: The slope is the change in *y*-values over the change in *x*-values. The change in *y*-values between successive rows is +2. The change in *x*-values is +1. Thus the slope is $\frac{+2}{+1}$ or 2. Then plugging the first row into $y = mx + b$ yields $1 = 2(-5) + b$ or $1 = -10 + b$. Then, $b = 11$. The equation is $y = 2x + 11$. Therefore, choice B is correct.

35. A: The slope and the *y*-intercept can be read from the graph. The *y*-intercept is −3. The slope is 2. The equation of this line is $y = 2x - 3$. Therefore, choice A is correct.

36. A: Substituting $y = 6$ and $x = 3$ into $y = kx$ yields $6 = k(3)$ which means $k = 2$. The equation for this direct variation is $y = 2x$. Therefore, choice A is correct.

37. C: Parallel lines have the same slope. Since the slope of $y = 3x - 1$ is 3, the slope of the parallel line is also 3. The equation can be found by plugging the slope and point into the formula $y - y_1 = m(x - x_1)$. Then, $y - 1 = 3(x + 2)$. Rearranging yields $y - 1 = 3x + 6$ or $y = 3x + 7$. Therefore, choice C is correct.

38. D: The slopes of perpendicular lines are negative multiplicative inverses or negative reciprocals. Since the slope of $y = 4x + 3$ is 4, the slope of a line perpendicular to it is $-\frac{1}{4}$. The equation of the line can be found by plugging the slope and point into the formula $y - y_1 = m(x - x_1)$. Then, $y + 6 = -\frac{1}{4}(x - 3)$ or $y + 6 = -\frac{x}{4} + \frac{3}{4}$. Finally, $y = -\frac{x}{4} - \frac{21}{4}$. Therefore, choice D is correct.

39. C: The slope and the point can be substituted into the formula $y - y_1 = m(x - x_1)$. Then $y - 5 = -3(x + 1)$. Rearranging yields $y - 5 = -3x - 3$ or $y = -3x + 2$. Therefore, choice C is correct.

40. D: The domain is all of the *x*-values for the function. The range is all *y*-values for the function. This function includes all real *x*-values, but *y* is always non-negative. The domain is $-\infty \leq x \leq +\infty$, and the range is $y \geq 0$. Therefore, choice D is correct.

41. B: The domain is all of the *x*-values for the function. The range is all *y*-values for the function. This function includes all real *x*-values, but the minimum *y*-value is 1. The domain is $-\infty \leq x \leq +\infty$, and the range is $y \geq 1$. Therefore, choice B is correct.

42. C: The graph of y_1 matches Figure 1. The graph opens up, with the vertex at (0,0), so it is the parent function of $y = x^2$. Figure 3 opens downward with the vertex at (0,0), so it is simply the parent function flipped: $y = -x^2$. Figure 4 opens upward with the vertex at (0,10), so it shifts the parent function up by 10: $y = x^2 + 10$. Figure 2 is not represented. It opens upward with the vertex at (0,–10), so the parent function is shifted down 10: $y = x^2 - 10$. Therefore, choice C is correct.

43. The correct answer is 5. First write Equation A in slope-intercept form: $y = mx + b$ where m is the slope and b is the y-intercept.

$$\begin{aligned} 5y - 100x &= 25 \\ 5y &= 100x + 25 \\ y &= 20x + 5 \end{aligned}$$

Based on the slope-intercept form of Equation A, the y-intercept is $b = 5$.

44. The correct answer is –16. To solve for A, we first need to multiply each side by 5: $6 + A = (-2)(5) \rightarrow 6 + A = -10$. Then subtract 6 from each side to find that $A = -16$.

45. The correct answer is –5. The solution follows:

$$\begin{aligned} x^2 + 10x &= -25 \\ x^2 + 10x + 25 &= 0 \\ (x + 5)(x + 5) &= 0 \\ x &= -5 \end{aligned}$$

46. The correct answer is –3. The axis of symmetry is the central line about which the parabola is symmetric. The vertex is the point at which the parabola intersects the axis of symmetry (which is also the parabola's maximum or minimum point). The equation of the axis of symmetry is $x = -3$.

47. The correct answer is $x = 5$ or $x = -2$. The equation $3x^2 - 9x - 30 = 0$ can be factored to $3(x^2 - 3x - 10) = 0$ which can be further factored to $3(x - 5)(x + 2) = 0$. Then $x - 5 = 0$ or $x + 2 = 0$ which means $x = 5$ or $x = -2$.

48. A: The equation of the parabola $y = 3(x - 1)^2 + 4$ is in vertex form. This can be expanded to $y = 3(x^2 - 2x + 1) + 4$ or $y = 3x^2 - 6x + 3 + 4$. Finally, this can be written as $y = 3x^2 - 6x + 7$. Therefore, choice A is correct.

49. D: The equation of the parabola $y = -2(x + 5)^2 - 8$ is in vertex form. This can be expanded to $y = -2(x^2 + 10x + 25) - 8$ or $y = -2x^2 - 20x - 50 - 8$. Finally, this can be written as $y = -2x^2 - 20x - 58$. Therefore, choice D is correct.

50. D: Since the rock drops from a 50-foot cliff, the vertex is at (0, 50). Since at 1 second after the fall, the height of the rock is 34 feet, the point (1, 34) also satisfies the function. The vertex form of a parabola is $y = a(x - h)^2 + k$ in which (h, k) is the vertex. Substituting in vertex (0, 50) for (h, k) and point (1, 34) for (x, y), $34 = a(1 - 0)^2 + 50$. Then $34 = a + 50$, and $a = -16$. The vertex form of this parabola is $y = -16x^2 + 50$ or $h = -16t^2 + 50$. Therefore, choice D is correct.

51. A: The zeros are the x-values for which the equation equals zero. If a quadratic equation can be factored, the zeros can be determined directly from the linear factors, by setting each factor equal to zero. Therefore, choice A is correct.

52. B: The zeros are the x-values where the graph of the quadratic equation crosses the x-axis. The graph crosses the x-axis at (–6, 0) and (1, 0). The zeros are –6 and 1. Therefore, choice B is correct.

53. C: This can be solved using the quadratic formula $x = \frac{-b \pm \sqrt{b^2-4ac}}{2a}$. Then, $x = \frac{-(-1) \pm \sqrt{(-1)^2-4(2)(-5)}}{2(2)}$. Simplifying, $x = \frac{1 \pm \sqrt{41}}{4}$. Therefore, choice C is correct.

54. D: This is an exponential growth function in the form of $y = ab^x$ where $a > 0$ and $b > 1$. Exponential growth functions have domains of all real values or $-\infty < x < +\infty$, and ranges of all positive real values or $0 < y < +\infty$. Therefore, choice D is correct.

55. B: This is an exponential decay function in the form of $y = ab^x$ where $a > 0$ and $0 < b < 1$. Exponential decay functions have domains of all real values or $-\infty \leq x \leq +\infty$, and ranges of all positive real values or $0 < y < +\infty$. Therefore, choice B is correct.

56. A: The equation is an exponential decay function in the form of $y = ab^x$ where $a > 0$ and $0 < b < 1$. Linear functions only have exponents of 1. Quadratic functions have at least one term of degree 2. Neither linear functions nor quadratic functions have a variable as an exponent. An exponential growth function is in the form of $y = ab^x$ where $a > 0$ and $b > 1$. Therefore, choice A is correct.

57. C: This is an exponential growth function in the form of $y = ab^x$. In this equation, a is the initial value, which is given here as 20,000, and b is the growth rate. An annual increase of 2% means a rate of growth of 102%, or 1.02. Then, $y = 20{,}000(1.02)^x$. Therefore, choice C is correct.

58. D: This is an exponential growth function in the form of $y = ab^x$. In this equation, a is the initial value, which is given here as 500, and b is the growth rate. An interest rate of 4% means a rate of growth of 104%, or 1.04. Then, $y = 500(1.04)^x$. Therefore, choice D is correct.

59. B: The equation is an exponential decay function in the form of $y = ab^x$ where $a > 0$ and $0 < b < 1$. Since the truck loses 10% of its value every year, each year its value is 90% of the previous year. That means $b = 0.90$. The starting value, or a, is given here as 40,000. The equation is $y = 40{,}000(0.90)^x$. Therefore, choice B is correct.

60. A: Use a graphing calculator. Press STAT and select EDIT to enter the x-values into List 1. Enter the y-values into List 2. To perform the regression, press STAT and select CALC. Scroll down to ExpReg. Hit ENTER. (Different graphing calculators might involve slightly different procedures; consult the instructions to your calculator.) The equation is $y = 57.22(0.870)^x$. Therefore, choice A is correct.

How to Overcome Test Anxiety

Just the thought of taking a test is enough to make most people a little nervous. A test is an important event that can have a long-term impact on your future, so it's important to take it seriously and it's natural to feel anxious about performing well. But just because anxiety is normal, that doesn't mean that it's helpful in test taking, or that you should simply accept it as part of your life. Anxiety can have a variety of effects. These effects can be mild, like making you feel slightly nervous, or severe, like blocking your ability to focus or remember even a simple detail.

If you experience test anxiety—whether severe or mild—it's important to know how to beat it. To discover this, first you need to understand what causes test anxiety.

Causes of Test Anxiety

While we often think of anxiety as an uncontrollable emotional state, it can actually be caused by simple, practical things. One of the most common causes of test anxiety is that a person does not feel adequately prepared for their test. This feeling can be the result of many different issues such as poor study habits or lack of organization, but the most common culprit is time management. Starting to study too late, failing to organize your study time to cover all of the material, or being distracted while you study will mean that you're not well prepared for the test. This may lead to cramming the night before, which will cause you to be physically and mentally exhausted for the test. Poor time management also contributes to feelings of stress, fear, and hopelessness as you realize you are not well prepared but don't know what to do about it.

Other times, test anxiety is not related to your preparation for the test but comes from unresolved fear. This may be a past failure on a test, or poor performance on tests in general. It may come from comparing yourself to others who seem to be performing better or from the stress of living up to expectations. Anxiety may be driven by fears of the future—how failure on this test would affect your educational and career goals. These fears are often completely irrational, but they can still negatively impact your test performance.

Elements of Test Anxiety

As mentioned earlier, test anxiety is considered to be an emotional state, but it has physical and mental components as well. Sometimes you may not even realize that you are suffering from test anxiety until you notice the physical symptoms. These can include trembling hands, rapid heartbeat, sweating, nausea, and tense muscles. Extreme anxiety may lead to fainting or vomiting. Obviously, any of these symptoms can have a negative impact on testing. It is important to recognize them as soon as they begin to occur so that you can address the problem before it damages your performance.

The mental components of test anxiety include trouble focusing and inability to remember learned information. During a test, your mind is on high alert, which can help you recall information and stay focused for an extended period of time. However, anxiety interferes with your mind's natural processes, causing you to blank out, even on the questions you know well. The strain of testing during anxiety makes it difficult to stay focused, especially on a test that may take several hours. Extreme anxiety can take a huge mental toll, making it difficult not only to recall test information but even to understand the test questions or pull your thoughts together.

Effects of Test Anxiety

Test anxiety is like a disease—if left untreated, it will get progressively worse. Anxiety leads to poor performance, and this reinforces the feelings of fear and failure, which in turn lead to poor performances on subsequent tests. It can grow from a mild nervousness to a crippling condition. If allowed to progress, test anxiety can have a big impact on your schooling, and consequently on your future.

Test anxiety can spread to other parts of your life. Anxiety on tests can become anxiety in any stressful situation, and blanking on a test can turn into panicking in a job situation. But fortunately, you don't have to let anxiety rule your testing and determine your grades. There are a number of relatively simple steps you can take to move past anxiety and function normally on a test and in the rest of life.

Physical Steps for Beating Test Anxiety

While test anxiety is a serious problem, the good news is that it can be overcome. It doesn't have to control your ability to think and remember information. While it may take time, you can begin taking steps today to beat anxiety.

Just as your first hint that you may be struggling with anxiety comes from the physical symptoms, the first step to treating it is also physical. Rest is crucial for having a clear, strong mind. If you are tired, it is much easier to give in to anxiety. But if you establish good sleep habits, your body and mind will be ready to perform optimally, without the strain of exhaustion. Additionally, sleeping well helps you to retain information better, so you're more likely to recall the answers when you see the test questions.

Getting good sleep means more than going to bed on time. It's important to allow your brain time to relax. Take study breaks from time to time so it doesn't get overworked, and don't study right before bed. Take time to rest your mind before trying to rest your body, or you may find it difficult to fall asleep.

Along with sleep, other aspects of physical health are important in preparing for a test. Good nutrition is vital for good brain function. Sugary foods and drinks may give a burst of energy but this burst is followed by a crash, both physically and emotionally. Instead, fuel your body with protein and vitamin-rich foods.

Also, drink plenty of water. Dehydration can lead to headaches and exhaustion, especially if your brain is already under stress from the rigors of the test. Particularly if your test is a long one, drink water during the breaks. And if possible, take an energy-boosting snack to eat between sections.

Along with sleep and diet, a third important part of physical health is exercise. Maintaining a steady workout schedule is helpful, but even taking 5-minute study breaks to walk can help get your blood pumping faster and clear your head. Exercise also releases endorphins, which contribute to a positive feeling and can help combat test anxiety.

When you nurture your physical health, you are also contributing to your mental health. If your body is healthy, your mind is much more likely to be healthy as well. So take time to rest, nourish your body with healthy food and water, and get moving as much as possible. Taking these physical steps will make you stronger and more able to take the mental steps necessary to overcome test anxiety.

Mental Steps for Beating Test Anxiety

Working on the mental side of test anxiety can be more challenging, but as with the physical side, there are clear steps you can take to overcome it. As mentioned earlier, test anxiety often stems from lack of preparation, so the obvious solution is to prepare for the test. Effective studying may be the most important weapon you have for beating test anxiety, but you can and should employ several other mental tools to combat fear.

First, boost your confidence by reminding yourself of past success—tests or projects that you aced. If you're putting as much effort into preparing for this test as you did for those, there's no reason you should expect to fail here. Work hard to prepare; then trust your preparation.

Second, surround yourself with encouraging people. It can be helpful to find a study group, but be sure that the people you're around will encourage a positive attitude. If you spend time with others who are anxious or cynical, this will only contribute to your own anxiety. Look for others who are motivated to study hard from a desire to succeed, not from a fear of failure.

Third, reward yourself. A test is physically and mentally tiring, even without anxiety, and it can be helpful to have something to look forward to. Plan an activity following the test, regardless of the outcome, such as going to a movie or getting ice cream.

When you are taking the test, if you find yourself beginning to feel anxious, remind yourself that you know the material. Visualize successfully completing the test. Then take a few deep, relaxing breaths and return to it. Work through the questions carefully but with confidence, knowing that you are capable of succeeding.

Developing a healthy mental approach to test taking will also aid in other areas of life. Test anxiety affects more than just the actual test—it can be damaging to your mental health and even contribute to depression. It's important to beat test anxiety before it becomes a problem for more than testing.

Study Strategy

Being prepared for the test is necessary to combat anxiety, but what does being prepared look like? You may study for hours on end and still not feel prepared. What you need is a strategy for test prep. The next few pages outline our recommended steps to help you plan out and conquer the challenge of preparation.

Step 1: Scope Out the Test

Learn everything you can about the format (multiple choice, essay, etc.) and what will be on the test. Gather any study materials, course outlines, or sample exams that may be available. Not only will this help you to prepare, but knowing what to expect can help to alleviate test anxiety.

Step 2: Map Out the Material

Look through the textbook or study guide and make note of how many chapters or sections it has. Then divide these over the time you have. For example, if a book has 15 chapters and you have five days to study, you need to cover three chapters each day. Even better, if you have the time, leave an extra day at the end for overall review after you have gone through the material in depth.

If time is limited, you may need to prioritize the material. Look through it and make note of which sections you think you already have a good grasp on, and which need review. While you are studying, skim quickly through the familiar sections and take more time on the challenging parts.

Write out your plan so you don't get lost as you go. Having a written plan also helps you feel more in control of the study, so anxiety is less likely to arise from feeling overwhelmed at the amount to cover.

STEP 3: GATHER YOUR TOOLS

Decide what study method works best for you. Do you prefer to highlight in the book as you study and then go back over the highlighted portions? Or do you type out notes of the important information? Or is it helpful to make flashcards that you can carry with you? Assemble the pens, index cards, highlighters, post-it notes, and any other materials you may need so you won't be distracted by getting up to find things while you study.

If you're having a hard time retaining the information or organizing your notes, experiment with different methods. For example, try color-coding by subject with colored pens, highlighters, or post-it notes. If you learn better by hearing, try recording yourself reading your notes so you can listen while in the car, working out, or simply sitting at your desk. Ask a friend to quiz you from your flashcards, or try teaching someone the material to solidify it in your mind.

STEP 4: CREATE YOUR ENVIRONMENT

It's important to avoid distractions while you study. This includes both the obvious distractions like visitors and the subtle distractions like an uncomfortable chair (or a too-comfortable couch that makes you want to fall asleep). Set up the best study environment possible: good lighting and a comfortable work area. If background music helps you focus, you may want to turn it on, but otherwise keep the room quiet. If you are using a computer to take notes, be sure you don't have any other windows open, especially applications like social media, games, or anything else that could distract you. Silence your phone and turn off notifications. Be sure to keep water close by so you stay hydrated while you study (but avoid unhealthy drinks and snacks).

Also, take into account the best time of day to study. Are you freshest first thing in the morning? Try to set aside some time then to work through the material. Is your mind clearer in the afternoon or evening? Schedule your study session then. Another method is to study at the same time of day that you will take the test, so that your brain gets used to working on the material at that time and will be ready to focus at test time.

STEP 5: STUDY!

Once you have done all the study preparation, it's time to settle into the actual studying. Sit down, take a few moments to settle your mind so you can focus, and begin to follow your study plan. Don't give in to distractions or let yourself procrastinate. This is your time to prepare so you'll be ready to fearlessly approach the test. Make the most of the time and stay focused.

Of course, you don't want to burn out. If you study too long you may find that you're not retaining the information very well. Take regular study breaks. For example, taking five minutes out of every hour to walk briskly, breathing deeply and swinging your arms, can help your mind stay fresh.

As you get to the end of each chapter or section, it's a good idea to do a quick review. Remind yourself of what you learned and work on any difficult parts. When you feel that you've mastered the material, move on to the next part. At the end of your study session, briefly skim through your notes again.

But while review is helpful, cramming last minute is NOT. If at all possible, work ahead so that you won't need to fit all your study into the last day. Cramming overloads your brain with more information than it can process and retain, and your tired mind may struggle to recall even previously learned information when it is overwhelmed with last-minute study. Also, the urgent nature of cramming and the stress placed on your brain contribute to anxiety. You'll be more likely to go to the test feeling unprepared and having trouble thinking clearly.

So don't cram, and don't stay up late before the test, even just to review your notes at a leisurely pace. Your brain needs rest more than it needs to go over the information again. In fact, plan to finish your studies by noon or early afternoon the day before the test. Give your brain the rest of the day to relax or focus on other things, and get a good night's sleep. Then you will be fresh for the test and better able to recall what you've studied.

Step 6: Take a Practice Test

Many courses offer sample tests, either online or in the study materials. This is an excellent resource to check whether you have mastered the material, as well as to prepare for the test format and environment.

Check the test format ahead of time: the number of questions, the type (multiple choice, free response, etc.), and the time limit. Then create a plan for working through them. For example, if you have 30 minutes to take a 60-question test, your limit is 30 seconds per question. Spend less time on the questions you know well so that you can take more time on the difficult ones.

If you have time to take several practice tests, take the first one open book, with no time limit. Work through the questions at your own pace and make sure you fully understand them. Gradually work up to taking a test under test conditions: sit at a desk with all study materials put away and set a timer. Pace yourself to make sure you finish the test with time to spare and go back to check your answers if you have time.

After each test, check your answers. On the questions you missed, be sure you understand why you missed them. Did you misread the question (tests can use tricky wording)? Did you forget the information? Or was it something you hadn't learned? Go back and study any shaky areas that the practice tests reveal.

Taking these tests not only helps with your grade, but also aids in combating test anxiety. If you're already used to the test conditions, you're less likely to worry about it, and working through tests until you're scoring well gives you a confidence boost. Go through the practice tests until you feel comfortable, and then you can go into the test knowing that you're ready for it.

Test Tips

On test day, you should be confident, knowing that you've prepared well and are ready to answer the questions. But aside from preparation, there are several test day strategies you can employ to maximize your performance.

First, as stated before, get a good night's sleep the night before the test (and for several nights before that, if possible). Go into the test with a fresh, alert mind rather than staying up late to study.

Try not to change too much about your normal routine on the day of the test. It's important to eat a nutritious breakfast, but if you normally don't eat breakfast at all, consider eating just a protein bar. If you're a coffee drinker, go ahead and have your normal coffee. Just make sure you time it so that the caffeine doesn't wear off right in the middle of your test. Avoid sugary beverages, and drink enough water to stay hydrated but not so much that you need a restroom break 10 minutes into the test. If your test isn't first thing in the morning, consider going for a walk or doing a light workout before the test to get your blood flowing.

Allow yourself enough time to get ready, and leave for the test with plenty of time to spare so you won't have the anxiety of scrambling to arrive in time. Another reason to be early is to select a good seat. It's helpful to sit away from doors and windows, which can be distracting. Find a good seat, get out your supplies, and settle your mind before the test begins.

When the test begins, start by going over the instructions carefully, even if you already know what to expect. Make sure you avoid any careless mistakes by following the directions.

Then begin working through the questions, pacing yourself as you've practiced. If you're not sure on an answer, don't spend too much time on it, and don't let it shake your confidence. Either skip it and come back later, or eliminate as many wrong answers as possible and guess among the remaining ones. Don't dwell on these questions as you continue—put them out of your mind and focus on what lies ahead.

Be sure to read all of the answer choices, even if you're sure the first one is the right answer. Sometimes you'll find a better one if you keep reading. But don't second-guess yourself if you do immediately know the answer. Your gut instinct is usually right. Don't let test anxiety rob you of the information you know.

If you have time at the end of the test (and if the test format allows), go back and review your answers. Be cautious about changing any, since your first instinct tends to be correct, but make sure you didn't misread any of the questions or accidentally mark the wrong answer choice. Look over any you skipped and make an educated guess.

At the end, leave the test feeling confident. You've done your best, so don't waste time worrying about your performance or wishing you could change anything. Instead, celebrate the successful completion of this test. And finally, use this test to learn how to deal with anxiety even better next time.

Review Video: Test Anxiety
Visit mometrix.com/academy and enter code: 100340

Important Qualification

Not all anxiety is created equal. If your test anxiety is causing major issues in your life beyond the classroom or testing center, or if you are experiencing troubling physical symptoms related to your anxiety, it may be a sign of a serious physiological or psychological condition. If this sounds like your situation, we strongly encourage you to seek professional help.

Online Resources

Due to our efforts to try to keep this book to a manageable length, we've created a link that will give you access to all of your online resources:

mometrix.com/resources719/cchsalgebra

It's Your Moment, Let's Celebrate It!

Share your story @mometrixtestpreparation